Theoretical Biology

Series Editor

Yoh Iwasa, Kyushu University, Fukuoka, Japan

The "Theoretical Biology" series publishes volumes on all aspects of life sciences research for which a mathematical or computational approach can offer the appropriate methods to deepen our knowledge and insight.

Topics covered include: cell and molecular biology, genetics, developmental biology, evolutionary biology, behavior sciences, species diversity, population ecology, chronobiology, bioinformatics, immunology, neuroscience, agricultural science, and medicine.

The main focus of the series is on the biological phenomena whereas mathematics or informatics contribute the adequate tools. Target audience is researchers and graduate students in biology and other related areas who are interested in using mathematical techniques or computer simulations to understand biological processes and mathematicians who want to learn what are the questions biologists like to know using diverse mathematical tools.

Atsushi Mochizuki

Structural Theories for Dynamical Networks in Biology

Springer

Atsushi Mochizuki
Institute for Life and Medical Sciences
Kyoto University
Kyoto, Japan

ISSN 2522-0438 ISSN 2522-0446 (electronic)
Theoretical Biology
ISBN 978-981-96-6025-4 ISBN 978-981-96-6026-1 (eBook)
https://doi.org/10.1007/978-981-96-6026-1

This Springer imprint is published by the registered company Springer Nature Singapore Pte Ltd.
The registered company address is: 152 Beach Road, #21-01/04 Gateway East, Singapore 189721, Singapore

Preface

From the middle of the twentieth century to the present day, the development of life sciences has been remarkable, and the momentum shows no signs of slowing down. Today, even the functions of life, including those of humans, are no longer mysterious, and are being understood through the concrete mechanisms involving biomolecules such as genes and proteins. Based on our understanding of biomolecules, research and development of treatments for diseases that were once thought to be incurable is progressing. While life sciences continue to make great strides, research fields that do not rely on conventional experimental methods such as molecular biology and biochemistry are attracting attention as a new trend. One such field is mathematical biology or theoretical biology, which uses mathematics and computer simulations to try to elucidate biological phenomena.

As the life sciences have developed, it has become clear that many biomolecules are involved in various biological phenomena, and that the dynamics of the complex systems consisting of interactions between biomolecules are the essence of biological functions. Mathematical methods are indispensable for providing an understanding of the dynamics in relation to the information on biomolecular interactions obtained experimentally. In addition, computational methods are expected to be useful in providing an integrated understanding of the accumulated information on biomolecules.

These theoretical methods are actually being used in the field of life science research, and are becoming established as a major method. In many cases of joint research between experiment and theory, mathematical models are constructed for newly elucidated molecular mechanisms, and the dynamics are confirmed. There are also cases where theoretical predictions of unknown biomolecular regulation are made, and then experimentally verified. In this way, a cycle of theoretical prediction and experimental verification is beginning to take shape. It could be said that a new life science is beginning to unfold, in which experiments and theory complement each other and advance together.

In addition, theoretical biology is not just about applying the methods developed in the fields of mathematics and physics to biological phenomena. In the process of trying to elucidate biological phenomena, new mathematical methods are being developed to solve problems specific to life, and it is showing its own unique development as a theoretical science.

This book was written with the aim of introducing the theoretical biology, which continues to grow with great expectations, in an easy-to-understand way, using specific topics, and at the same time explaining the main mathematical methods. Below, I will list some of the features of this book.

Structure of the Book

Each chapter covers an important topic that is currently attracting attention in the field of biological research, and through the story of how biological phenomena are theoretically elucidated, the importance and interest of mathematical methods are naturally conveyed. At the same time, at the end of each chapter, there is a section on "Mathematical Methods" that summarizes the important mathematical methods used in each chapter so that they can be studied independently of the main text. In addition, there are exercises at the end of some chapters.

In the first half, we focus on the dynamics that arise from the regulation of biomolecular activity, such as gene expression and protein modification. We will look at some of the biological phenomena that arise from the dynamics of biomolecular activity, such as gene switches and the asymmetrical gene expression in mammals (Chap. 3), the circadian rhythms in *Drosophila* and segmentation clocks in mammals (Chap. 4), and the oscillations of reaction systems that do not involve transcription or translation (Chap. 5). By constructing differential equation models for the activity of biomolecules, we can understand biological behavior in terms of the behavior of dynamical systems, that is, the diversity of solutions and periodic oscillatory solutions. The book introduces the ideas of theoretical biology, and also provides a simple explanation of the basic theoretical analysis of dynamical systems and an overview of matrix theory. The "Mathematical Methods" at the end of each chapter are related to the topics of each chapter, but at the same time, they are arranged in a way that allows the reader to gradually progress from easy content to more advanced content, making it easy for individuals to learn.

In the second half of the book, we approach the understanding of the dynamics of complex systems composed of many biomolecules. Complex networks are characteristics of biological phenomena, and for example, in Chaps. 6 and 7, we focus on the gene regulatory network that includes 92 factors that govern cell fate specification, and in Chaps. 8 and 9, we focus on the reaction network that consists of 25 metabolic molecules and 45 reactions. For such complex systems, i.e., high-dimensional dynamical systems, it is difficult to gain an accurate understanding using only conventional methods based on mathematical models. Therefore, in this book, we introduce three "structural theories" that the author and his colleagues have developed in recent years as new theoretical methods. Structural theories are mathematical theories that determine important aspects of the behavior of dynamical systems from network information alone. They are also called model-free theories because they do not assume any model. Not only are they new as mathematical theories, but they are also practical and powerful for elucidating biological systems. I will introduce examples in which powerful predictions were derived using the structural theories for several complex networks, including the example mentioned above, and where further progress was made in elucidating the systems by coupling with experimental verifications.

How to Read This Book

The structure of this book has been designed to allow for a variety of different ways of reading it. The most common way of reading it would be to read the main text and then read the "Mathematical Methods" section to deepen your understanding, but you are not obliged to follow this approach. You could also read the main text only and get an idea of the way theoretical biology is thought about. Alternatively, you could read the "Mathematical Methods" section only and then work through the exercises to help you learn. Each chapter is independent of the others, so you can basically start reading from anywhere. If you already understand the content of the "Mathematical Methods," you can skip it. However, if you are reading "Mathematical Methods" in Section 2~5, it is better to read them in order from the beginning. The mathematical methods in Chaps. 7 and 8, which introduce the mathematics of the structural theories, may be a little difficult. If you can't understand everything at once, there is no need to worry. If you are interested in these theories, even if you don't fully understand the mathematics behind them, I would encourage you to give them a try. To understand this book, it is sufficient to have studied basic mathematics such as linear algebra and differential and integral calculus, and no other special mathematical knowledge is required. However, if you read this book with a textbook on dynamical systems or differential equation theory by your side, your understanding will deepen even further.

Intended Audience

The intended readers of this book are (1) students and young researchers with a background in theoretical sciences such as mathematical science and physics who are interested in biological phenomena, and (2) students and young researchers with a background in biological sciences who want to learn theoretical methods. It seemed difficult to attract the interest of both biological science and theoretical science students and researchers, but I thought that this could be achieved by writing about content that the author himself finds interesting.

This book also has the characteristic of being the first textbook in a language other than Japanese to introduce the "structural theories" in a comprehensive manner. Although the theories were developed with the aim of elucidating biological systems, they have attracted the attention of many theoretical researchers because of their potential to be applied to various complex systems beyond biological systems. Therefore, (3) theoretical researchers who are not limited to the field of life sciences and who are interested in structural theories are also assumed to be readers of this book.

I would be delighted if readers who pick up this book find it interesting. I believe that this book will provide readers with new perspectives and useful information. I also wrote this book with that hope. I would be extremely happy if this book satisfied readers' intellectual curiosity and contributed in some way to their happiness as scientists.

Kyoto, Japan Atsushi Mochizuki

Contents

Chapter 1
Introduction to Living Systems

Abstract As a beginning of this book, I will give an overview of biological systems, the mechanisms behind biological phenomena. I will provide a perspective from mathematical biology on the basic content of the life sciences and make drastic simplifications and reorganizations. In particular, this would be an introduction to biology for readers from theoretical fields. I hope that readers will feel more familiar with the complex biological phenomena.

Keywords Life sciences · Network system · Reaction system · Regulatory system · Graph theory

1.1 Characteristic Behaviors of Life and Living Systems

The purpose of biological sciences is to clarify the mechanisms by which the characteristic behaviors of living organisms are realized. The term "life science" is sometimes used to refer to the study of biological phenomena in the fields of medicine, science, agriculture, pharmacy, and engineering, without distinction among these fields. There is no better way to learn about the intriguing behaviors inherent to life than by referring to textbooks in biology and the life sciences (Alberts et al. 2017; Berg et al. 2019). Here are three general properties that the author feels are particularly interesting.

1. Homeostasis: Cells take in organic materials, using some parts of them to build their internal structures and extract chemical energy, and expelling the rest. As the inflow and outflow continue, the materials constituting the cell should be replaced with time; however, the intracellular structure is maintained, and the cell functions continue continuously. It is notable that the structure and functions are maintained almost unchanged while the substances as an entity are replaced. In addition, cellular functions remain nearly invariant under widely fluctuating temperature and nutritional conditions without being affected much. This property is called homeostasis.
2. Development and morphogenesis: In many multicellular organisms, complex body structures with biological functions are created through a process called

A. Mochizuki, *Structural Theories for Dynamical Networks in Biology*, Theoretical Biology, https://doi.org/10.1007/978-981-96-6026-1_1

development. A single fertilized egg undergoes repeated cell division and differentiation to produce cells with diverse properties, which are appropriately positioned in the three-dimensional structure of the organism. It can be said that the functions of the three-dimensional body of an organism are realized by the appropriate spatial arrangement of cells with different properties. The basic structure of the adult body does not differ greatly among individuals, and many developmental processes are considered to be genetically determined. For this reason, developmental genetics, which elucidates developmental phenomena through genetic manipulation, has made great progress since the latter half of the twentieth century (Gilbert 2020). On the other hand, it is difficult to say that the developmental process has been fully elucidated from the viewpoint of how three-dimensional morphology and structure are embedded in a one-dimensional DNA sequence.

3. Autonomy: The behaviors of life, such as homeostasis and development, appear to be well-designed and well-controlled systems. Whereas, these behaviors are not realized by an observer outside the system who monitors the system and controls it according to the situation. For example, in the early developmental stage of the protoderm, the embryo undergoes highly dynamic deformation. This is not achieved by an external force but by the autonomous deformation of a large number of cells, each of which deforms appropriately according to its position in the embryo. In other words, each cell knows its own position in the cell population and behaves appropriately depending on the position.

Life sciences have elucidated the mechanisms of biomolecules that give rise to these dynamic behaviors. It seems that there are aspects that have been greatly elucidated and aspects that have not been so well understood. One aspect that has greatly advanced is the understanding of the functions of biomolecules, such as genes and proteins, and the relationships between them, such as reactions and regulations. Information on the relationships between biomolecules has been summarized as reaction systems and regulatory systems. In the following, we will review the understanding of reaction and regulatory systems as one of the major achievements of the life sciences.

1.2 Chemical Reaction Systems and Reaction Networks

The various phenomena occurring in a cell are essentially all chemical reactions. In other words, biological phenomena are realized by the accumulation of chemical reactions. Chemical reaction equations are a common way to describe state transitions caused by chemical reactions. The substance before the reaction (reactant) and the substance after the reaction (product) are connected by arrows, and if there is more than one reactant or product, they are connected by a plus sign (+).

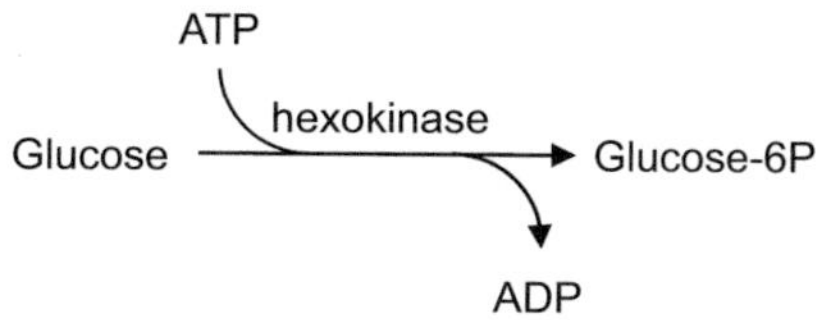

Fig. 1.1 Graphical representation of a chemical reaction (1.1)

$$\text{Glucose} + \text{ATP} \longrightarrow \text{Glucose-6P} + \text{ADP} \tag{1.1}$$

Equation (1.1) is one of the reactions in the metabolic system and is the first reaction in the process where glucose is broken down into pyruvate (glycolytic system).

Chemical reaction equations can be represented graphically, as shown in Fig. 1.1. Sometimes the enzyme that catalyzes the reaction is written alongside the arrow. If there is more than one reactant or product, the arrow has multiple start points or multiple endpoints, respectively. If chemical substances are distinguished based on their elemental composition, in principle at least one of the starting points or endpoints of the arrow should be multiple. Whereas, when describing chemical reactions in living organisms, it is sometimes the case that the abundant substances such as water molecules, ions, ATP, etc. may not be explicitly shown, and as a result, a reaction may be represented as an unbranched arrow. Although I used the term "chemical reaction," Eq. (1.1) and Fig. 1.1 can be generally used to describe state transitions.

If we write the rate of the chemical reaction shown in Eq. (1.1) and Fig. 1.1 as a function of time t, $r_{HK}(t)$, then the temporal changes in the concentration of glucose, $u_{Glc}(t)$, and glucose-6-phosphate, $u_{G6P}(t)$, based on this reaction, follow the differential equation:

$$\frac{d}{dt}u_{Glc}(t) = -r_{HK}(t) \tag{1.2a}$$

$$\frac{d}{dt}u_{G6P}(t) = r_{HK}(t) \tag{1.2b}$$

For more information on the dynamics of chemical reaction systems, see Chaps. 5 and 8.

Reactions in living organisms have been identified in a variety of methods. Most of the reactions in living organisms are catalyzed by proteins called enzymes. Once an enzyme has been identified, the substrates and products of the reactions it catalyzes have been biochemically determined. On the other hand, it is also possible to use information in databases to estimate the function of an enzyme based on other enzymes with similar gene sequences. Among the reactions that occur in organisms, the metabolic system is a typical example where the correspondence between reactants, products, and enzymes has been well studied.

Reactions in living organisms usually do not occur in isolation but rather are connected in a chain in which the products of one reaction become the reactants of

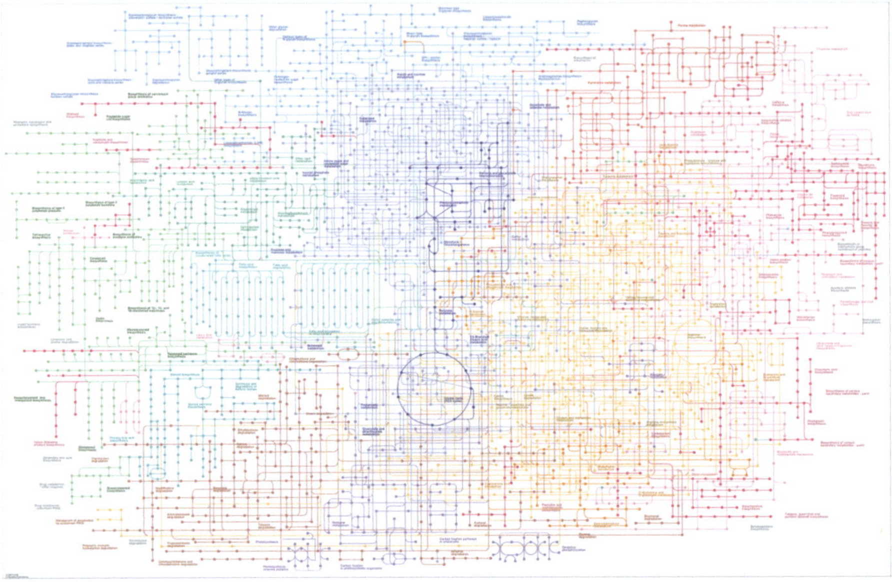

Fig. 1.2 In vivo chemical reaction network. (KEGG: Kyoto Encyclopedia of Genes and Genomes) https://www.genome.jp/kegg/

another reaction. A graphical representation of a chain of reactions where products and reactants are shared is called a "chemical reaction network." In this book, I use the terms "system" and "network" almost in the same sense, with the latter being used to emphasize size in particular.

Figure 1.2 shows a network of reactions identified in various organisms and summarized in KEGG, a well-known database. There is an argument that to understand the dynamics of chemical reactions in living organisms accurately, one must consider the dynamics of such a whole system. However, Fig. 1.2 is too large to analyze, so researchers usually select a portion of the system and focus on the behavior of the subsystem (see, for example, Fig. 1.1 and Fig. 8 in Chap. 8). The mathematical theory introduced in Chap. 8 guarantees that at least some aspects of the behavior can be studied exactly by subsystems extracted from a chemical reaction system under a certain rule.

1.3 Gene Regulatory Systems and Regulatory Networks

Regulatory systems describe the relationships between biomolecules focusing on a different aspect from reaction systems. Reaction systems describe the "state transitions" of biomolecules caused by reactions, whereas regulatory systems describe the "cause-and-effect" relationships among the activities of biomolecules. A typical example is a gene regulatory system.

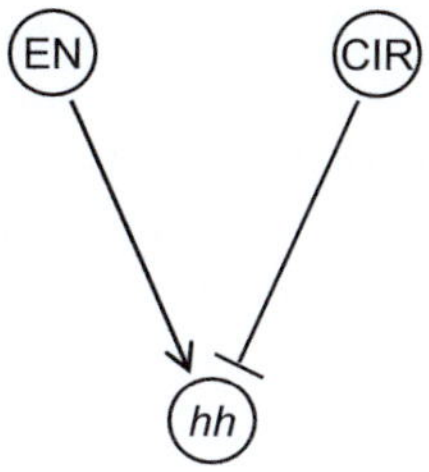

Fig. 1.3 Graphical representation of gene regulation. Transcriptional regulation of the Hedgehog gene (*hh*) in the *Drosophila* segmentation. Engrailed gene transcript (EN) activates the transcription, and cubitus interruptus repressor (CIR) represses the transcription

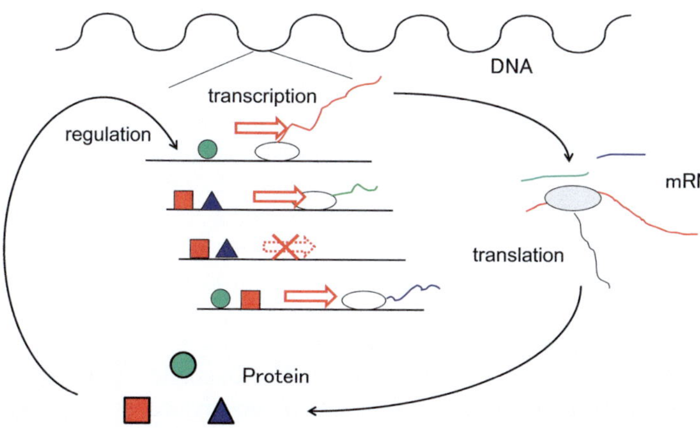

Fig. 1.4 Transcriptional regulation of genes. Proteins are produced from activated genes via mRNA. Some of the proteins are involved in gene transcription regulation as transcription factors

Figure 1.3 shows the transcription regulation of the hedgehog gene by two different transcription factors in the *Drosophila* segmentation. The upper two circles represent protein transcription factors, and the lower circle represents the gene (or mRNA as a result of transcription). Arrows indicate activation of transcription, and barheads indicate repression of transcription. This type of representation, with the vertices representing biomolecules and the arrows (edges) representing the regulatory relationships between them, is called a regulatory system (Fig. 1.4).

From genes activated by transcription regulation, mRNA containing the sequence information of DNA is produced. According to the sequence information in the mRNA, a protein, a macromolecule consisting of 20 different amino acids, is produced. Proteins have various functions, such as (i) catalyzing chemical reactions as enzymes, (ii) producing mechanical force as motor proteins, or producing structures as the intracellular skeleton. Another important function is as (iii) a transcription factor. Transcription factors are proteins that switch between active and inactive transcription by binding to the vicinity of gene sequences on DNA. The proteins produced as a result of gene activation regulate the activity/inactivity of some genes again, and the transcription factors form a regulatory network (Fig. 1.1 in Chap. 6 and Fig. 1.1 in Chap. 7). Gene transcription may not be closed within a single cell but is also regulated by signaling molecules (ligands) from outside the cell. In addition to transcription factors, chemical modifications of DNA molecules and histones (proteins that bind to DNA) are also involved in transcriptional regulation, which is called epigenetic regulation.

To consider dynamics based on a regulation system such as the one shown in Fig. 1.3, each vertex is assigned a variable, such as the amount of mRNA or protein. In the case of Fig. 1.3, the differential equation describing the dynamics of the amount of mRNA in the hedgehog gene can generally be written as

$$\frac{d}{dt}u_{hh}(t) = F_{hh}(u_{hh}(t); v_{EN}(t), v_{CIR}(t)). \tag{1.3}$$

See Chap. 7 for details. u_{hh} represents the amount of hedgehog gene mRNA, and v_{EN} and v_{CIR} represent the amount of EN and CIR proteins, respectively.

F_{hh} is a function that determines the temporal change of u_{hh} and is the most important piece of information. From the regulation relationship shown in Fig. 1.3, we can see that F_{hh} is an increasing function of v_{EN} and a decreasing function of v_{CIR}.

In biology, gene regulatory relationships have been identified in various ways. For example, regulatory relationships can be inferred from binding information, such as which transcription factors bind to the regulatory region of a gene of interest. The most reliable method is using perturbation experiments to identify the effect of each factor on the activities of (the other) genes. For example, if the experimental disruption of the function of gene A results in inactivated gene B, which was originally active, it can be inferred that gene A originally activated gene B. A prominent example of identifying the regulatory relationships among more than 90 genes by repeats of perturbation experiments is the network system governing the cell fate specification of the ascidian, which will be discussed in Chaps. 6 and 7.

Regulatory systems are used to describe causal relationships among various biomolecules in general, not only genes. For example, regulatory factors are not limited to transcription factors but can also include epigenetic factors. For protein phosphorylation, the relationship between a kinase protein and a substrate protein may be expressed as a regulatory system. Furthermore, there is an idea of phenomenological regulatory systems, which may include not only biomolecules but also non-material factors (e.g., cellular behaviors, environmental conditions, etc.).

Figure 1.5 shows the regulatory network of the signal transduction pathway in which the binding of epidermal growth factors (EGF) to receptors on the cell membrane leads to the sequential activation of various proteins and small molecules within the cell and eventually changes in gene activity. Secreted proteins like EGF are known to induce various cell behaviors such as cell proliferation, cell migration, cell differentiation, and cell death, which are shown at the bottom vertices of Fig. 1.5. This figure shows the summary of the information obtained from a large number of the published papers in a single map. As mentioned above, cellular responses induced by signal transduction include cell migration, cell differentiation,

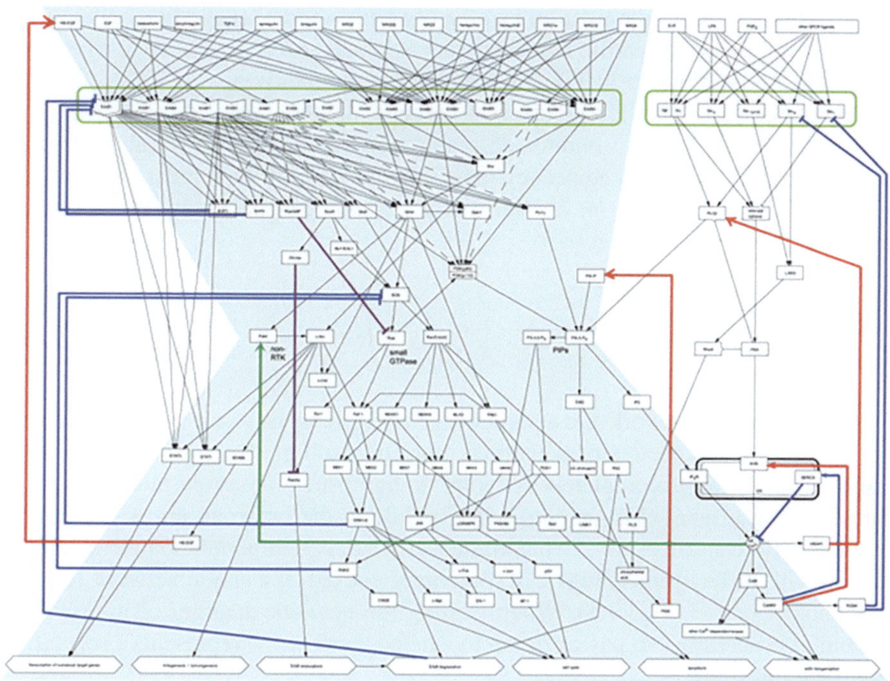

Fig. 1.5 Regulatory network of the signal transduction pathway downstream of the EGF receptor and its family. (Oda et al. 2005)

and oncogenesis, which have been studied in different fields or contexts. Figure 1.5 is meant to provide a comprehensive viewpoint by integrating information that has been studied in different contexts.

The following are definitions in graph theory related to regulatory systems. In graph theory, a structure consisting of a set of vertices V and a set of edges E is called a graph $\Gamma = (V, E)$, where each edge in E is assumed to connect two vertices (which may be identical) in V. When each edge in a graph Γ has a direction and distinguishes a starting point (tail) and an ending point (head), it is called a directed graph (see also Chap. 7). Regulatory networks almost conform to the definition of directed graphs, but as mentioned above, they often distinguish between activating and inhibitory regulations by arrows and barheads. Even in the context of regulatory networks in biology, arrows are sometimes used to indicate just causality, without distinguishing between activation and inhibition, which matches the definition of directed graphs. At first glance, a reaction network also seems to consist of a set of edges connecting vertices and could be called a directed graph. However, a directed edge in graph theory is defined to connect a single tail and a single head, and a reaction network, which can have multiple starting or ending points, does not fit the definition of a graph.

In addition to the representation by a set of vertices and a set of edges, a graph can also be described by a matrix: Consider an $n \times n$ integer matrix $A = (a_{ij})$, where n is the number of vertices ($n = |V|$). A component of the matrix $a_{ij} = 1$ if there exists a directed edge $(j \rightarrow i) \in E$ between two vertices $i, j \in V$, and $a_{ij} = 0$ if there is no directed edge $(j \rightarrow i)$. An undirected graph is represented as a symmetric matrix. Such a matrix is called an adjacency matrix. In the case of the regulation system in Fig. 1.3, if EN, CIR, and *hh* are vertices 1, 2, and 3, respectively, the adjacency matrix of this graph is given by

$$A = \begin{pmatrix} 0 & 0 & 0 \\ 0 & 0 & 0 \\ 1 & 1 & 0 \end{pmatrix}. \tag{1.4}$$

Both a reaction network and a regulatory network consist of a set of vertices and a set of arrows, and their representations may be similar. Note, however, that the meaning of arrows is completely different. In the former, an arrow indicates a state transition, i.e., "what changes into what," while in the latter, an arrow indicates a regulation (or causality), i.e., "what influences what. It is important to note that these arrows with different meanings can sometimes coexist in a single network (e.g., in this book, Chap. 3, Fig. 1.5). When reading the network diagram of a biological system, it is necessary to pay attention to whether each arrow represents a regulation or a state transition.

Information on reactions and regulations has been accumulated as data throughout the history of life science, and the network information that summarizes these data has gradually become more complex. It is possible that new molecular mechanisms will be discovered based on information about reactions or regulation discovered in the past. Data is also important in the sense that new discoveries made by individual researchers can be placed in the context of past knowledge systems. Life science is increasingly becoming a data science, and recently, the term "Big Data of Life" has been used to describe it.

1.4 Toward Understanding Dynamics

While biology and life sciences have increased our knowledge of biomolecules and their interactions, there are still some aspects that have not been fully explored. One such aspect is our understanding of temporal changes or dynamics. As discussed in Sect. 1.1, the fascination and depth of biological phenomena lie in their dynamic behavior. While much progress has been made in analyzing the mechanisms of biomolecules, there is still room for clarification of the dynamic behaviors themselves and the principles that give rise to such behaviors.

The importance of dynamics is also highly recognized in experimental life sciences, and measurement techniques to capture the dynamics of biomolecules are rapidly developing. However, it is still difficult to capture the dynamics of many molecules simultaneously with high temporal resolution.

The main objective of this book is to introduce the methods in mathematical and theoretical sciences to understand the dynamics of biological systems from their network information. Understanding the dynamics based on information about molecular interactions, such as reaction systems and regulation systems, should strongly promote our understanding of biological phenomena. For this purpose, mathematical and theoretical methods should be more commonly used in biology and life sciences.

In other fields of natural science, temporal change continues to be an important issue and has formed a field called "dynamical systems theory" in mathematics and physics. This book also relies greatly on the benefits of dynamical systems theory. On the other hand, there are several problems unique to the life sciences that do not exist in mathematics and physics and which make it difficult to understand the dynamics of living systems.

First, there is still large limitation in the species of molecules whose dynamics can be directly measured and in the conditions under which such measurements can be made. As mentioned above, the simultaneous and high-resolution measurement of the dynamics of multiple molecular species is a critical issue in the current life sciences. In other words, even if one builds a mathematical model of a living system and calculates its detailed dynamics computationally, the opportunity of quantitatively validating it and judging the correctness of the mathematical model is currently limited. Current experimental life sciences are sometimes based on the hypothetical dynamics of biomolecules that are not actually measured. Experimental researchers also seem to face the frustration of not being able to measure dynamics directly. Whereas, at the same time, this leads to high expectations for contributions by mathematical and theoretical methods.

Second, in most cases, the system size is huge. For example, it is estimated that the human genome contains about 22,000 genes. In other words, to capture the activity changes of all human genes at the same time, a system with at least tens of thousands of variables must be considered. Since it is impossible to understand the whole picture of such a system, we usually study subsystems that contain only the biomolecules that are considered important in experimental biology. Even then, it is usually a large network. For example, in the early development of the ascidian, which will be discussed in Chap. 7, the gene regulatory network governing the fates of seven tissues is determined and includes 92 factors.

Mathematical analysis of large-scale dynamical systems has been difficult in the mathematical sciences and physics, and the main strategies are either to approximate the system with a small number of variables or to give up on mathematical analysis and just perform numerical simulations. In addition, mathematical analysis, which has been the forte of the mathematical sciences up to now, does not always yield useful results in the life sciences. For example, in the life sciences, it is important to understand the dynamics of each specific system, and analyses such as the average behavior when the network structures are randomly changed do not contribute much to the elucidation of the actual life system (although they give theoretical meaning).

Third, the information is incomplete. Not only the limitation of observing dynamics, but most living systems provide only partial information to determine their dynamics. For example, from the regulatory system in Fig. 1.3, we can read off the function F_{hh} on the right-hand side of the differential Eq. (1.3) as an increasing function of v_{EN} and a decreasing function of v_{CIR}, but we do not know anything more than that. In order to construct the dynamics in concrete detail, we must build a mathematical model under the specific assumption of the functional form of F_{hh}. For small systems, the use of mathematical models is effective, but for large network systems, it means that the dynamics can be determined only after making a large number of assumptions about the functions and parameter values. There are some efforts to quantitatively determine transcription functions and translation functions experimentally, but even determining them for a single gene requires a great deal of effort. In short, although a large amount of information on biomolecular interactions has been accumulated in the modern life sciences, important aspects of information for determining dynamics are missing.

These are not necessarily challenges that need to be overcome by experimental biologists. There may be clever ways to use mathematical methods to avoid or solve these problems. In some cases, new mathematical methods may arise from these difficulties or the needs of the life sciences. This book introduces a variety of mathematical and theoretical methods, always keeping in mind the above issues and considering how to contribute to life sciences. In some chapters, these issues are directly addressed. As a result, newly developed mathematical methods will be introduced and proposed.

1.5 Summary

Graphical representations of reaction systems or regulatory systems have been actively used by experimental researchers since the dawn of the life sciences. The reason for this may be that graphs were useful for understanding the biological systems in which multiple biomolecules are intertwined. In addition, it seems to suggest that there has been an inherent desire to understand biological phenomena as dynamics since the birth of the life sciences. The purpose of this book is to meet such long-standing expectations in the field of life sciences.

References

Alberts B, Johnson A, Lewis J, Morgan D, Raff M, Roberts K (2017) Molecular biology of the cell, 6th edn. Garland Science, p 1464

Berg JM, Tymoczko JL, Gatto GJ Jr, Stryer L (2019) Biochemistry, 9th edn. WH Freeman, p 1208

Gilbert SF (2020) Developmental biology, 12th edn. Sinauer Associates Inc, p 792

Oda K, Matsuoka Y, Funahashi A, Kitano H (2005) A comprehensive pathway map of epidermal growth factor receptor signaling. Mol Syst Biol 1:1–17

Chapter 2
One-Dimensional Dynamical Systems

Abstract Various biological functions originate from temporal changes in the activity of biomolecules, especially genes. In this chapter, we will consider how biomolecules exhibit temporal changes under the regulations of activation and repression. Mathematical methods for dealing with temporal changes, or dynamics, have been developed in the field of dynamical systems theory, and a variety of analytical methods are available. In order to understand the basic concepts of mathematical analysis, this chapter first focuses on the activity of a single molecule and then considers its dynamics.

Keywords Differential equation · Hill function · Equilibrium point · Stable equilibrium point · Unstable equilibrium point · Linear approximation

2.1 Dynamics of Synthesis and Degradation of Biomolecules

Let us consider the dynamics of gene expression under transcription regulation. The expression or activity level of a gene is the amount of biomolecules such as mRNA or protein. Let u be the concentration of the mRNA of interest, and let the temporal change of u be given by a differential equation

$$\frac{du}{dt} = a - bu. \tag{2.1}$$

The first term a on the right side is the rate of increase of mRNA, i.e., the rate of transcription, where transcription at a constant rate is assumed. The second term bu represents the rate of decrease of mRNA, i.e., the decay term. Since an activity level is a concentration of a biomolecule, the term for decrease in concentration due to inactivation or degradation is always necessary. The decay term can generally take various forms as an increasing function of concentration, but here it is simply assumed to be proportional to the concentration.

The general solution $u(t)$ satisfying the differential Eq. (2.1) is

A. Mochizuki, *Structural Theories for Dynamical Networks in Biology*, Theoretical Biology, https://doi.org/10.1007/978-981-96-6026-1_2

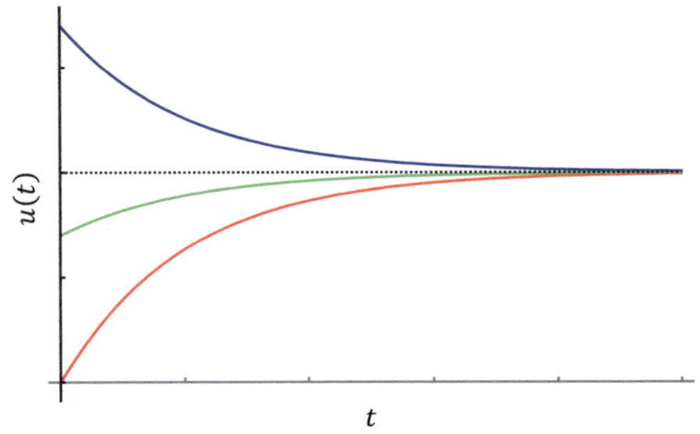

Fig. 2.1 Trajectories of solutions of (2.1) for different initial conditions

$$u(t) = \frac{a}{b} + Ce^{-bt}, \tag{2.2}$$

where C is an arbitrary constant derived from the integral (Exercise 2.1). Since the differential Eq. (2.1) gives only the local change of the variable $u(t)$ along the time axis, the solution $u(t)$ is not uniquely determined by itself. By giving $u(0) = u_0$ as the concentration of mRNA at time $t = 0$, Eq. (2.2) can also be written as

$$u(t) = \frac{a}{b} + \left(u_0 - \frac{a}{b}\right)e^{-bt}. \tag{2.3}$$

The concentration of mRNA at time $t = 0$, $u(0) = u_0$, is called the initial condition. A solution with a specific value for an arbitrary constant is called a particular solution. For example, when $u(0) = 0$ is given as the initial condition, the solution $u(t)$ is

$$u(t) = \frac{a}{b}\left(1 - e^{-bt}\right). \tag{2.4}$$

Finding a solution that satisfies both a differential equation and an initial condition is called an initial value problem. In general, the behaviors of a solution $u(t)$ of a differential equation depend on the initial conditions. For example, under the initial condition of $u_0 < a/b$, $u(t)$ of solution (2.3), it will show an upward convex curve that increases with time and approaches a/b. However, if the initial condition is given as $u_0 > a/b$, the trajectory of $u(t)$ will be an upward concave curve (Fig. 2.1).

In Fig. 2.1, $u(t)$ appears to approach a certain value with the progress of time. Substituting $u = a/b$ into (2.1) here satisfies $du/dt = 0$, indicating that the temporal change of u is zero at exactly this value. Such a state point is called an equilibrium. In this model, without depending on the initial state $u(0) = 0$, the gene activity level $u(t)$ will approach the equilibrium point a/b after a sufficient time has passed.

For a more nontrivial example, let us consider the case in which the transcription rate is not constant but changes by regulation. As illustrated in Fig. 2.2, assume a

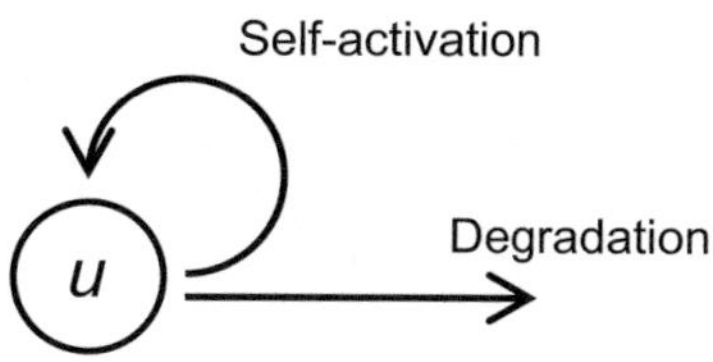

Fig. 2.2 A gene that regulates the activity of itself

circuit in which the protein product of the gene acts as a transcription factor to promote its own activity.

Although it is possible to distinguish between mRNA and protein concentrations and express them by different variables (e.g., u and v), for the sake of simplicity in this chapter, we use the one-variable differential equation (Griffith 1968)

$$\frac{du}{dt} = f(u) - bu, \tag{2.5}$$

where f is an increasing function of u representing the transcription rate, and bu represents the decay rate as in (2.1).

Here, for the transcription function f, we would like to use a function that starts from near 0 at small value of u, rapidly increases at a certain value of u, and then saturates and approaches its maximum value as u increases further. Such a typical example is the Hill function (Hill, 1910)

$$H(u; K) = \frac{u^n}{K^n + u^n}. \tag{2.6}$$

The Hill function was proposed to model the cooperative effect of multiple ligand-binding receptors, where the binding of one ligand enhances the binding of the second and third ligands. Consider a receptor with n ligand-bindings at maximum, and the binding-dissociation reactions are:

$$\begin{aligned} &L + R \rightleftarrows LR \\ &L + LR \rightleftarrows L_2R \\ &\vdots \\ &L + L_{n-1}R \rightleftarrows L_nR \end{aligned} \tag{2.7}$$

where L and R are the ligands and the receptor molecule, respectively. If we consider the case that once one ligand binds, the affinity of subsequent ligands is extremely high, the binding-dissociation reaction between the receptor and the ligands can be approximated by the reversible reaction equation

$$nL + R \rightleftarrows L_nR. \tag{2.8}$$

The dissociation constant is given by

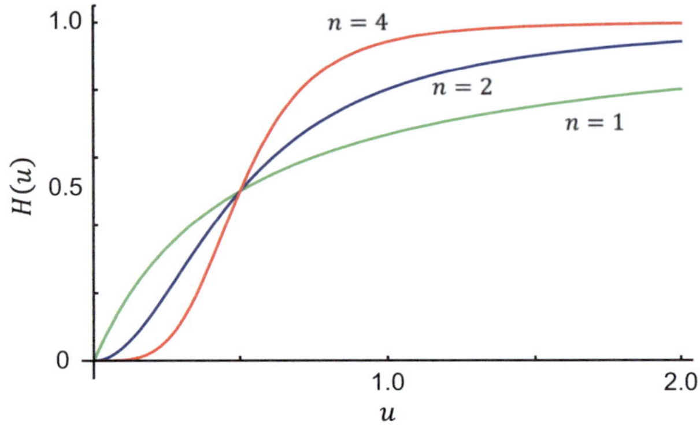

Fig. 2.3 Hill function for $K = 0.5$ and $n = 1$, 2, 4. When $n > 1$, it appears as a S-shaped increasing function. The larger n is the more rapidly the value changes around the threshold K

$$K_d^n = \frac{[L]^n[R]}{[L_nR]}, \tag{2.9}$$

where [] is the concentration of the molecule or receptor. The ratio of binding receptors can be calculated using (2.9) in the form of the Hill function below

$$\frac{[L_nR]}{[R]+[L_nR]} = \frac{[L]^n}{K_d^n+[L]^n}. \tag{2.10}$$

The Hill function $H(u; K)$ is nearly 0 when u is small, but rapidly increases around $u = K$ and approaches 1. The value $u = K$ around which the activity switches is sometimes called the threshold value. The number n is called the Hill coefficient, and when this coefficient is large, the value of H increases rapidly around the threshold value. In general, a function that is not simply proportional to a variable is called a nonlinear function. The Hill function is a typical nonlinear function in biology and is often used to describe switch-like behavior or S-shaped functions in various biological phenomena, not only ligand binding (Fig. 2.3).

Using the Hill function with Hill coefficient of 2 for the transcription function in Eq. (2.5), let us consider the following differential equation

$$\frac{du}{dt} = \frac{u^2}{K^2+u^2} - bu, \tag{2.11}$$

where b and K are positive constants. There are two possible interpretations for using the Hill function for a transcription function. One is that multiple transcription factors can bind to the transcriptional regulatory region, and there is a cooperative effect in the binding, and the other is that the Hill function is used for convenience to express the non-linearity of transcriptional regulation.

Equation (2.11) is not easy to solve because the regulatory function is nonlinear. Therefore, let us first calculate the dynamics numerically. Numerical calculation is one of the most important and practical methods for mathematical models of

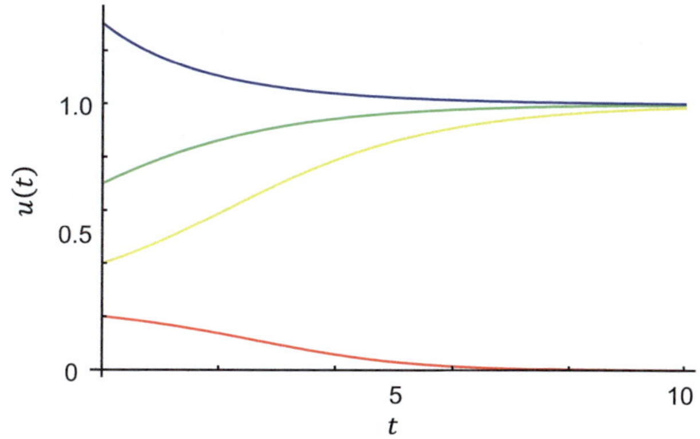

Fig. 2.4 Solution trajectory of the discretized formulation (2.13) of Eq. (2.11), solved numerically. Where $K = 0.5$ and $b = 0.8$. Various initial conditions were given, and the time evolution was computed for each

biological phenomena because mathematical models in the phenomena often include nonlinear terms. In Eq. (2.11), $u(t)$ is a quantity that changes continuously with time. Let $k = 0, 1, \cdots$ be discrete points on the time axis at intervals of Δt, and let u_k be the value of u at the k th point, then the discrete equation for u_k is given by

$$\frac{u_{k+1} - u_k}{\Delta t} = \frac{u_k^2}{K^2 + u_k^2} - bu_k. \tag{2.12}$$

Equation (2.12) is an approximation of Eq. (2.11) when Δt is sufficiently small. Reorganizing Eq. (2.12) with respect to k yields equation

$$u_{k+1} = u_k + \Delta t\left(\frac{u_k^2}{K^2 + u_k^2} - bu_k\right). \tag{2.13}$$

By Eq. (2.13), the value u_{k+1} at time $t = (k + 1)\Delta t$ is determined explicitly from the value u_k at time $t = k\Delta t$. There are several ways to derive the discrete form from the continuous equation. The method of discretization in the time direction as in (2.13) is the simplest, called the Euler method.

Figure 2.4 shows the time evolution of Eq. (2.11) with various initial conditions. Unlike Eq. (2.2), there are multiple final destinations, and where they end up depends on the initial conditions. We would like to analyze this phenomenon in more detail.

2.2 Analysis of 1D Dynamical Systems

The initial value problem for differential equations is not always solvable analytically. However, if our aim is to understand the dynamical behavior approximately, it is sufficient to draw the graph of du/dt without using advanced mathematics. Let $F(u)$ denotes the right-hand side of the differential Eq. (2.11) as

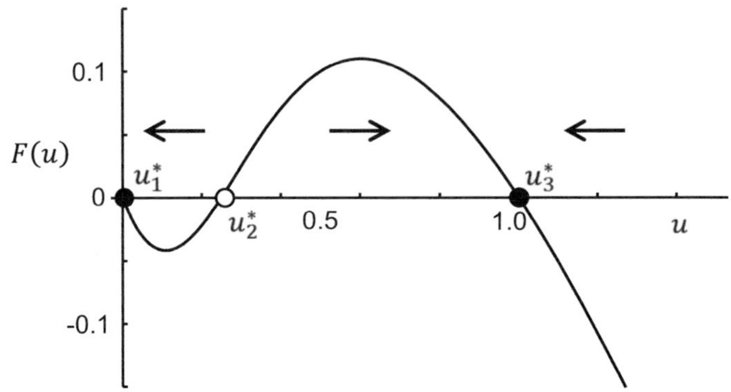

Fig. 2.5 $du/dt(=:F(u))$ of Eq. (2.14) as a function of u, where $K = 0.5$ and $c = 0.8$. The point where $F(u)$ intersects the u-axis ($F(u) = 0$) is the equilibrium point. The stability of an equilibrium point in 1-dimensional dynamics is determined by the slope of $F(u)$ at the point.

$$F(u) = \frac{u^2}{K^2 + u^2} - bu \tag{2.14}$$

shown in Fig. 2.5 as a function of u.

The direction of change in u is indicated by arrows: u increases when du/dt is positive and decreases when du/dt is negative. There are three points u_1^*, u_2^*, and u_3^*, that satisfy $F(u) = 0$, where u does not change with time, i.e., the equilibrium points. The behavior after a sufficient time depends on the initial condition $u(0) = u_0$: $u(t)$ approaches u_1^* if $u_0 < u_2^*$ and u_3^* if $u_2^* < u_0$. In general, when $F(u)$ is a continuous function, the direction of change of u (the sign of du/dt) is uniquely determined for each region of u divided by the equilibrium points.

Among these three equilibria u_1^*, u_2^*, and u_3^*, there are two types: one where $u(t)$ approaches from a point a short distance away and another where $u(t)$ moves away from a point a short distance away. The former (u_1^*, u_3^*) is called the stable equilibrium point and the latter (u_2^*) the unstable equilibrium point. From the graph, the stability of the equilibrium point is determined by the slope of $F(u)$ at the intersection with the horizontal axis. A point u^* that satisfies $F(u^*) = 0$ and $dF/du(u^*) < 0$ is a stable equilibrium point, and a point u^* that satisfies $F(u^*) = 0$ and $dF/du(u^*) > 0$ is an unstable equilibrium point. A system is sometimes called "bistable" when two stable equilibrium points appear, like u_1^* and u_3^* in this model.

Stability can be analyzed more quantitatively near the equilibrium point. Let $U := u - u^*$ be the small deviation of u from the equilibrium point u^*, and Taylor expansion of $F(u)$ for U gives

$$F(U + u^*) = F(u^*) + UF'(u^*) + \frac{U^2}{2} F''(u^*) + O(U^3)$$
$$\approx UF'(u^*) \tag{2.15}$$

Where we used $F(u^*) = 0$, and ignored terms of second order or higher for the small quantity U. Since $du/dt = dU/dt$, the differential equation for U near the equilibrium point is

$$\frac{dU}{dt} = F'(u^*)U. \tag{2.16}$$

The general solution of the linear ordinary differential Eq. (2.16) is

$$U(t) = Ce^{F'(u^*)t} \tag{2.17}$$

where C is the integral constant. (2.17) shows the dynamical behavior of the deviation $U(t)$ near the equilibrium point u^*. If $F'(u^*)$ is positive, the absolute value of the deviation $U(t)$ increases exponentially with time, that is, $u(t)$ moves away from the equilibrium point. If $F'(u^*)$ is negative, the absolute value of the deviation $U(t)$ decreases exponentially with time, that is, $u(t)$ approaches the equilibrium point. For an equilibrium point u^* satisfying $F(u^*) = 0$, it is an unstable if $F'(u^*) > 0$, and a stable if $F'(u^*) < 0$. In addition, $F'(u^*)$ also indicates the speed at which the equilibrium point is approaching or moving away from the equilibrium point. The method of analyzing the behavior near the equilibrium point by approximating it with linear ordinary differential equations is called "local stability analysis."

2.3 Summary

We have learned the basic concepts for understanding the dynamics of biomolecular activity, focusing on the activity of a single gene and analyzing its temporal changes by mathematical modeling using ordinary differential equations. Many actual biological systems contain multiple species of biomolecules. From the next chapter onward, we will consider multi-variable dynamical systems corresponding to them. What we have learned here is an important foundation and will be useful for understanding the next chapters and beyond.

Exercise

Exercise 2.1

For the system of ordinary differential equations with one variable (2.1), derive the general solution for $u(t)$. Also, derive the particular solution when the initial condition is $u(0) = 0$.

Exercise 2.2

The equilibrium points of the differential Eq. (2.11) are given by u^* satisfying $F(u^*) = 0$ in (2.14). Confirm that $u^* = 0$ is always an equilibrium point, and find the conditions for b and K such that there are three real equilibrium points.

Exercise 2.3

With regard to Eq. (2.14), show that $F'(u) = 2K^2u/(K^2 + u^2)^2 - b$, and show that the equilibrium point $u_1^* = 0$ is always stable.

Exercise 2.4

When the conditions for the existence of three equilibrium points obtained in Exercise 2.2 are satisfied, find the stability of the non-zero equilibrium points u_2^* and u_3^*. (Hint: You can simplify the stability condition by using the fact that u_2^* and u_3^* satisfy $F(u^*) = 0$, and then substitute u_2^* and u_3^* into the formula.

Exercise 2.5

Consider the differential equation model for the activity of a self-inhibiting gene, $du/dt = K^2/(K^2 + u^2) - bu$. What kind of change does $K^2/(K^2 + u^2)$ show with respect to u? Discuss the stability of the equilibrium point for this model.

References

Griffith JS (1968) Mathematics of cellular control processes. II. Positive feedback to one gene. J Theor Biol 20:209–216

Hill AV (1910) The possible effects of the aggregation of the molecules of hæmoglobin on its dissociation curves. J Physiol 40(Suppl):4–7

Chapter 3
Gene Regulatory Systems and Their Dynamics

Abstract A class of proteins called transcription factors has functions to activate or repress transcription of genes. In a cell, each of the many species of transcription factors regulates a number of target genes. At the same time, the amount of a transcription factor itself is regulated by transcription factors. Thus, gene regulations form a dynamical system as a whole. It is believed that various biological functions of the organism arise from the dynamics of gene activity of this system. In this chapter, we will learn the basics of mathematical methods to understand the dynamics of gene regulatory systems.

Keywords Gene regulation · Regulatory network · λ phage · Positive feedback · Nullcline analysis · Saddle point · Bistable · Saddle-node bifurcation · Left-right asymmetry · Local stability analysis

3.1 Dynamics of the Gene Regulatory System

Many biologists and life scientists now understand naturally that many biological functions arise from regulatory systems of biomolecules. Gene regulatory systems are a major research target in the life sciences, as they are the origin of biological functions. It is difficult to pinpoint exactly when the idea that "systems create functions" first arose in the history of biology and the life sciences. However, the discovery of the gene switch in λ phage in the early days of molecular biology is thought to be an important milestone in raising awareness of this idea.

The λ phage is a type of virus that infects bacteria. They exhibit two distinctly different behaviors within the host bacterial cell. In the "lysogenic cycle," λ phage genes are incorporated into the host chromosome and replicated along with the replication of the bacterial genome as if they were part of the host's genes. The other is called the "lytic cycle," in which phage DNA is actively transcribed and proteins are synthesized, and phage DNA is replicated at the same time. This cycle produces the next generation of phage one after another, eventually destroying the host bacterium. It has been revealed experimentally that these two extremely different phenotypes are realized by a simple gene regulatory mechanism (Fig. 3.1).

A. Mochizuki, *Structural Theories for Dynamical Networks in Biology*, Theoretical Biology, https://doi.org/10.1007/978-981-96-6026-1_3

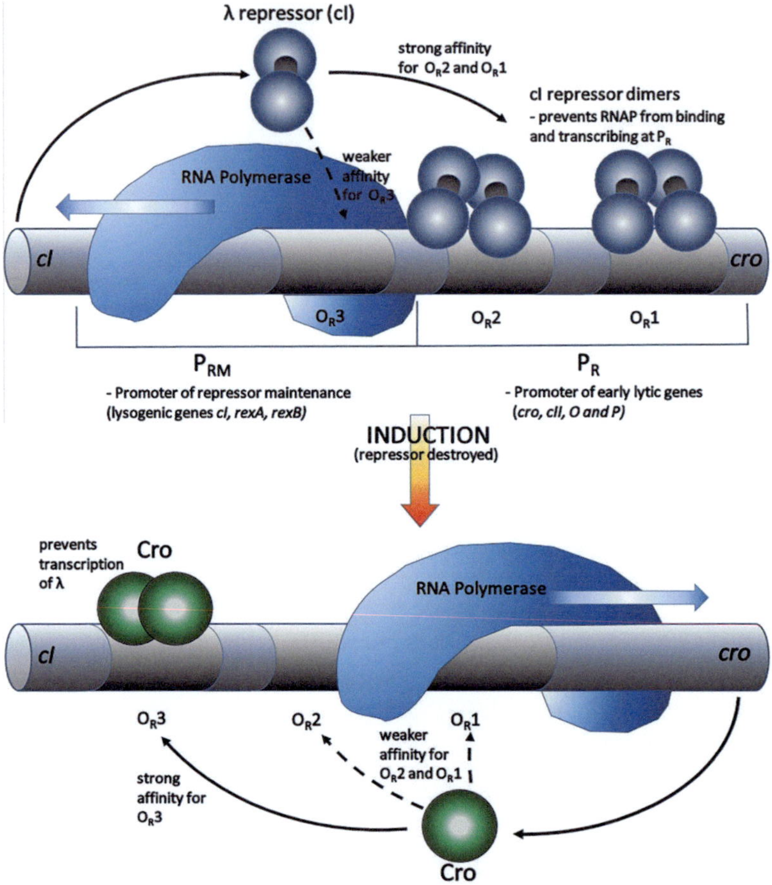

Fig. 3.1 Genetic switch of λ phage

Here is a brief summary of the regulation of the λ phage gene, focusing on the switch between lysogen and lytic (Ptashne 2004). The λ phage genes can be transcribed in two directions (leftward and rightward) from the transcriptional regulatory region. Both leftward and rightward transcription are regulated by a common binding region for transcription factors. When transcription proceeds in the leftward direction, the *cI* gene is transcribed downstream, and the protein of this transcript binds to the regulatory region as a factor to repress transcription in the rightward direction. On the other hand, when transcription proceeds in the rightward direction, the *cro* gene is activated downstream, and the protein of this product binds to the regulatory region to repress transcription in the leftward direction. When transcription proceeds in the rightward direction, proteins necessary for λ phage proliferation are synthesized downstream, whereas they are not synthesized when

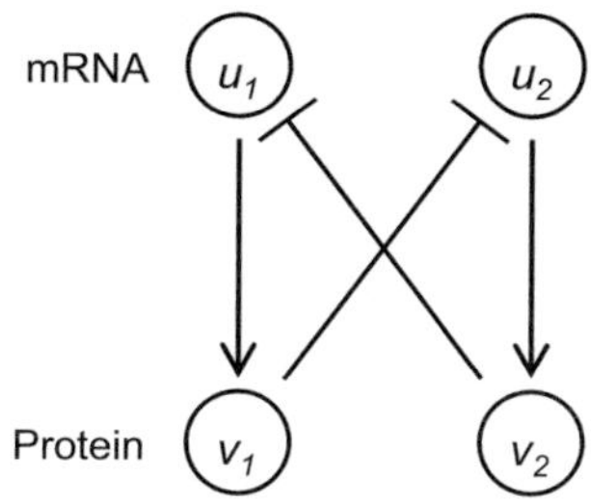

Fig. 3.2 Gene regulatory system of two genes mutually repressed. Here u_1 and u_2 are mRNA concentrations of *cI* and *cro* genes, respectively, and v_1 and v_2 are concentrations of CI and CRO proteins, respectively

transcription proceeds in the leftward direction. In other words, leftward and rightward transcription directly indicate the initiation of the lysogenic and lytic cycles, respectively.

Transcription of the *cI* gene and that of the *cro* gene are mutually repressive with their transcripts. Even without any bias in the early stages of the process, either only leftward or only rightward transcription may occur over time. This might be the mechanism that produces the extremely different phenotypes of lysogenic and lytic cycles.

Figure 3.2 shows a network of the regulation acting on the *cI* and *cro* genes of lambda phage. The upper nodes represent the mRNA concentrations of *cI* and *cro*, and the lower nodes represent the respective protein concentrations. In this figure, arrows represent activation regulation and barheads represent repression regulation. In the regulatory network shown in Fig. 3.2, if you start from a node and follow the direction indicated by the arrows and barheads, you will return to the original node. Such a closed path is called a feedback. (In graph theory, it is called a directed cycle.) Intuitively, it is expected that "the behavior of a factor will affect the original factor through a series of steps." If the original factor has a certain tendency (either high or low), and the feedback effect further promotes the tendency, it is called a positive feedback, and if the feedback effect reduces the original tendency, it is called a negative feedback. In the case of Fig. 3.2, there is a directed cycle $u_1 \to v_1 \to u_2 \to v_2 \to u_1$. Since there is an even number of inhibitions on its pathway, it seems to be a positive feedback intuitively.

We would like to discuss more precisely whether these intuitions are correct. Let us consider a mathematical model for the activity dynamics of the *cI* and *cro* genes (Arkin et al. 1998)

$$\frac{du_1}{dt} = \frac{a_1}{1 + (v_2/h_2)^n} - b_1 u_1 \tag{3.1a}$$

$$\frac{dv_1}{dt} = c_1 u_1 - d_1 v_1 \tag{3.1b}$$

$$\frac{du_2}{dt} = \frac{a_2}{1 + (v_1/h_1)^n} - b_2 u_2 \tag{3.1c}$$

$$\frac{dv_2}{dt} = c_2 u_2 - d_2 v_2 \tag{3.1d}$$

where u_1 and u_2 are the mRNA concentrations of the *cI* and *cro* genes, respectively, and v_1 and v_2 are the concentrations of the CI and CRO proteins, respectively, and both are assumed to be non-negative. Equations (3.1a) and (3.1c) represent the change in mRNA with time, with the first term representing an increase by transcription and the second term representing a decrease by degradation. Equations (3.1b) and (3.1c) represent the change in protein over time, with the first term representing an increase by translation and the second term representing a decrease by degradation. The parameters are all positive, where a_1 and a_2 are the maximum transcription rate, b_1 and b_2 are the degradation rate coefficients of mRNA, c_1 and c_2 are the translation rate coefficients, and d_1 and d_2 are the degradation rate coefficients of protein. For the transcription function, we consider a function that is rapidly suppressed when v_1 (v_2) exceeds the threshold value h_1 (h_2). Using the Hill function (Eq. 2.6) introduced in Chap. 2, we have $a_1\{1 - H(h_2; v_2)\}$ for a transcription function of Eq. (3.1a).

We want to know the dynamics of model (3.1), but it is difficult to handle it as it is. Therefore, we assume that both the synthesis and degradation of mRNA are sufficiently fast and that the dynamics of (3.1a) and (3.1c) are always at equilibrium ($du_1/dt = 0$, $du_2/dt = 0$). Substituting the obtained equations into (3.1b) and (3.1d), we obtain

$$\frac{dv_1}{dt} = \frac{a_1 c_1/b_1}{1 + (v_2/h_2)^n} - d_1 v_1 \tag{3.2a}$$

$$\frac{dv_2}{dt} = \frac{a_2 c_2/b_2}{1 + (v_1/h_1)^n} - d_2 v_2. \tag{3.2b}$$

Further, the number of parameters can be reduced by using a technique called nondimensionalization (Mathematical Method 3B). A system of differential equations

$$\frac{dv'_1}{dt'} = \frac{A_1}{1 + (v'_2)^n} - v'_1 \tag{3.3a}$$

$$\frac{dv'_2}{dt'} = \frac{A_2}{1 + (v'_1)^n} - Dv'_2 \tag{3.3b}$$

is obtained without losing the generality of the model.

In the following, we study the dynamics of $v'_1(t)$ and $v'_2(t)$ according to (3.3), where we ignore the prime ($'$). Even if a differential equation is given, its exact behavior is not always determined mathematically. On the other hand, in the case of a system of ordinary differential equations with two variables, the time evolution can be understood as a vector field $(dv_1/dt(v_1, v_2), dv_2/dt(v_1, v_2))$ on the plane (v_1, v_2).

The nullcline analysis is introduced as a graphical method to determine the approximate behavior of a system of ordinary differential equations with two variables. This is a two-dimensional version of the graphical method introduced in Chap. 2. First, curves satisfying $dv_1/dt = 0$ and $dv_2/dt = 0$ are drawn on the plane (v_1, v_2). These curves are called nullclines in the sense that the time derivative is zero on each of them. They are also called isoclines in the sense that "the slope $(dv_2/dt)/(dv_1/dt)$ on them is constant (zero or infinite)." The intersection of two nullclines $dv_1/dt = 0$ and $dv_2/dt = 0$ is an equilibrium point (or equilibrium state), where the values of (v_1, v_2) do not change by definition of the differential equation. The plane (v_1, v_2) is divided into several parts by nullclines. As long as the right-hand side of the differential equation is a continuous function, the sign of dv_1/dt (and dv_2/dt) is uniquely determined as either positive or negative for each region divided by the nullclines. By illustrating the direction of change of v_1 and v_2 in each region on the (v_1, v_2) plane with arrows, one can get an approximate behavior of the dynamics.

Figure 3.3 shows the nullclines and the trajectories of the numerical solutions under certain parameter conditions. In this case, the two nullclines $dv_1/dt = 0$ and $dv_2/dt = 0$ intersect at three points. That is, there are three equilibrium points $\{s_0, s_1, s_2\}$. From the graph, the dynamics depend on where initial state $(v_1(0), v_2(0))$ is located on the plane. If both CI and CRO protein concentrations are low $(v_1(0) \approx v_2(0) \approx 0)$ under the initial conditions, both proteins $v_1(t)$ and $v_2(t)$ increase once, but when $(v_1(t), v_2(t))$ crosses either of the nullclines, the direction of the trajectory changes and only one of $v_1(t)$ or $v_2(t)$ increases while the other decreases. Eventually, the equilibrium point is approached, either s_1 or s_2. The distance of $(v_1(t), v_2(t))$ from the equilibrium point decreases with time and converges to zero. If both protein concentrations are high initially $(v_1(0) \approx v_2(0) \approx 2)$, both CI and CRO initially decreases, but when $(v_1(t), v_2(t))$ crosses either of the nullclines, the direction changes and finally converge to either s_1 or s_2. The final destination between s_1 or s_2 is determined by the initial conditions, and the boundary on the (v_1, v_2) plane separating the final destination is called the separatrix.

Note the different properties between the three equilibrium points s_0, s_1, and s_2. Of these, s_1 and s_2 are the final dynamical endpoints, and from each surrounding region, $(v_1(t), v_2(t))$ approaches that point with time. On the other hand, from the neighbor of s_0, $(v_1(t), v_2(t))$ once approaches it with time but eventually changes direction and moves away from it. As mentioned in Chap. 2, an equilibrium point where the solution approaches from the surrounding area is called a stable equilibrium point. Conversely, an equilibrium point where the solution moves further away from the surrounding area is called an unstable equilibrium point. There are several types of unstable equilibrium points: unstable nodes where the solution moves away in all directions and saddle points where the solution approaches from one direction but moves away in another direction. Here, s_0 is a saddle point. Saddle points are not observed in one-dimensional dynamical systems but are observed in dynamical

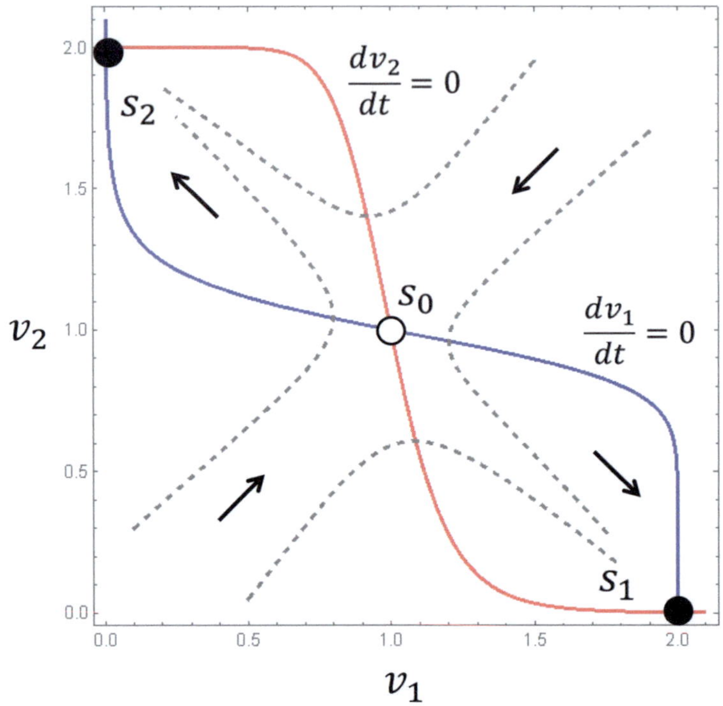

Fig. 3.3 Nullcline analysis for the dynamics (3.3). The blue and red curves show the nullclines of the CI protein concentration (v_1) and CRO protein concentration (v_2), respectively, and their intersections are the equilibrium points. The black and white circles indicate the stable and unstable equilibrium points, respectively. The gray dotted curves show the trajectories of the dynamics from four different initial conditions. The parameter values are $A_1 = A_2 = 2.0$, $D = 1.0$, $n = 10$

systems of two or more dimensions. The existence of two stable equilibrium points such as s_1 and s_2 in this model is called bistable.

Determining the stability of the equilibrium points is important for understanding dynamics. Although the direction of change at each point on the plane (v_1, v_2) can be determined by the nullcline analysis, it is not quantitative information and thus may not be sufficient to determine the stability of the equilibrium point. On the other hand, the behavior of the solution near the equilibrium point can be determined by a method called local stability analysis (3A mathematical method) even for complex dynamical systems. In the case of the system (3.3), it is confirmed that s_1 and s_2 are stable nodes, and s_0 is a saddle point by the local stability analysis as expected from Fig. 3.3. By combining the nullcline analysis with local stability analysis, the dynamic behavior of at least a two-dimensional dynamical system can be determined.

The nullcline analysis shown in Fig. 3.3 seems to capture the observed behavior of the *cI* gene and the *cro* gene during the infection process of the λ phage: either only the *cI* gene becomes active or only the *cro* gene becomes active, as a result of

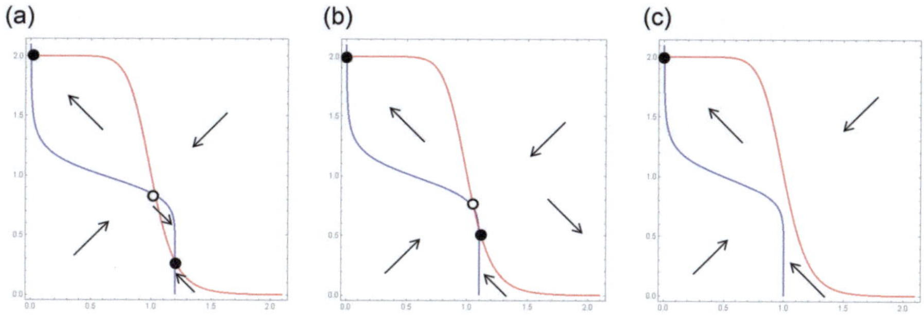

Fig. 3.4 Nullclines and equilibrium points under different parameter values. The parameter values are (**a**) $A_1 = 1.2$, (**b**) $A_1 = 1.1$, and (**c**) $A_1 = 1.0$. $A_2 = 2.0$, $D = 1.0$, and $n = 10$ are common to all three figures

mutual repression of the two genes. We can also see that there is no other stable state and that one of the two states is realized even if the initial condition is intermediate between the two. Which of the two states is realized is determined solely by the balance of the activities of the two genes at the initial condition. However, it should be noted that the behavior of this model depends on the values of the parameters, and they must be chosen appropriately for the expected behavior to emerge. An example is shown in Fig. 3.4.

Figure 3.4 shows the change in nullclines as the value of A_1 is decreased from those used in Fig. 3.3 in the order (a), (b), and (c). From (a) to (b), the stable equilibrium point s_1 and the saddle point s_0 gradually approach each other, and from (b) to (c), they fuse and disappear together. In this way, the dynamic behavior can change discontinuously even though the parameters are continuously changed. In dynamical systems, this phenomenon is called "bifurcation" and has been studied intensively. Depending on the type of changes in the dynamical behavior, bifurcations are classified into several types, such as pitchfork, Hopf, and so on. In the case of Fig. 3.4, as a node and a saddle point are fused and disappear, this type is called a saddle-node bifurcation. Bifurcations will be discussed again in subsequent chapters.

Bifurcation is not merely a mathematical concept; it often has important biological implications as well. In Fig. 3.4c, the dynamics after s_1 and s_0 have annihilated, there is only one stable equilibrium point, s_2. No matter what initial conditions are given for (v_1, v_2), the dynamics will eventually reach the equilibrium point s_2 where only v_2 is high. This indicates that if the transcription rate is extremely asymmetric, for example, if the transcription rate of the *cI* gene is less than half of that of the cro gene, the bistability due to mutual repression will not be realized.

What is shown here is just one of the possible behaviors that the model can exhibit, and it does not necessarily mean that this will happen in actual organisms. However, if we know how the behavior changes depending on the parameters, we can tell which coefficients are worth paying attention to in actual systems.

3.2 Systems for Asymmetric Gene Expression in Mammals

As a practical example of how a gene regulatory system can produce differences in cellular properties, let us discuss gene regulation that produces left-right asymmetry in early mammalian development. It has been shown that the first asymmetry to appear in mouse early embryos is a unidirectional flow of water from right to left, caused by the cilia rotating clockwise in the embryonic body surface structure called the node (Nonaka et al. 2002). As a result of receiving the flow signal, the first left-right asymmetry of gene expression in mouse development appears in about a dozen cells on side of the node (Fig. 3.5a). Two genes called Cerberus-like protein 2 (Cerl2) and Wnt3 (Wnt) play important roles in this phenomenon. Before nodal flow, Cerl2 gene expression is high and Wnt gene expression is low on both the left and right sides of the node. However, after nodal flow, gene expression changes only in the cells on the left side of the node, with low expression of the Cerl2 gene and high expression of the Wnt gene (Fig. 3.5a).

The group led by Professor Hiroshi Hamada (then of Osaka University, now of RIKEN Center for Biosystems Dynamics Research) has been working to elucidate the molecular mechanisms that create differences between left and right in mammals. As part of this project, they focused on the first left-right differences in gene expression and clarified the regulatory relationship between these genes (Nakamura et al. 2012) (Fig. 3.5b). In particular, they found that nodal flow signaling promotes the degradation of Cerl2 mRNA (regulation (2) in Fig. 3.5b). We would like to understand the dynamics of gene expression based on these regulations and whether they are the essential mechanism that produces the asymmetry in gene expression.

Nakamura et al. attempted to understand the dynamics of this system by constructing a mathematical model for the mRNA and protein concentrations of Cerl2 and Wnt, respectively (Nakamura et al. 2012). Here, we assume that the dynamics on the left and right sides of the node occur independently, but that the

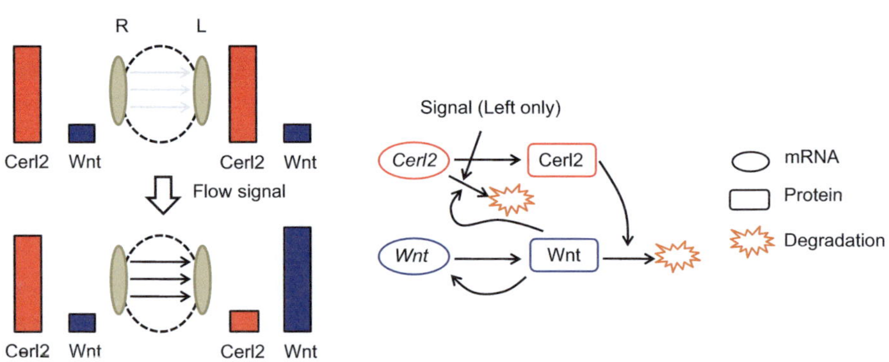

Fig. 3.5 Left-right asymmetry in gene expression in early mammalian development. (**a**) Changes in gene expression before and after nodal flow. (**b**) Two genes (Cerl2 and Wnt) are involved in the creation of left-right asymmetry and the regulation between them. (From Nakamura et al. 2012). (1), (2), and (3) are the three regulatory components of feedback. The details are explained in Sect. 3.4

mechanism is the same on both sides, except for the presence or absence of the node flow signal. In other words, the differences in gene expression between the left and right sides of the node are considered to be two states produced by a single system, depending on the presence or absence of nodal flow signals. In this way, the dynamics in the left and right sides are studied by a single system of ordinary differential equations:

$$\frac{du_c}{dt} = a_r - d_{cr}(S; v_w)u_c \tag{3.4a}$$

$$\frac{dv_c}{dt} = c_r u_c - d_{cp} v_c \tag{3.4b}$$

$$\frac{du_w}{dt} = f_w(v_w) - d_{wr} u_w \tag{3.4c}$$

$$\frac{dv_w}{dt} = c_w u_w - d_{wp}(v_c) v_w \tag{3.4d}$$

where u_c is the mRNA concentration of Cerl2 and v_c is the protein concentration of Cerl2, u_w is the mRNA concentration of Wnt, and v_w is the protein concentration of Wnt. S is the signal magnitude, which is 0 on the right side of the node and positive on the left side of the node, and is considered to increase with time. a_r, $d_{cr}(S; v_w)$, c_r, and d_{cp} are parameters or functions related to Cerl2 dynamics, which are the transcription rate, mRNA degradation rate coefficient, translation rate coefficient, and degradation rate coefficient of the protein. Similarly, $f_w(v_w)$, d_{wr}, c_w, and $d_{wp}(v_c)$ are parameters or functions related to the dynamics of Wnt and represent the transcription rate, mRNA degradation rate coefficient, translation rate coefficient, and protein degradation rate coefficient, respectively.

The degradation rate coefficient of Cerl2 mRNA, $d_{cr}(S; v_w)$, is an increasing function of the signal S and the Wnt protein v_w, the Wnt transcription function $f_w(v_w)$ is an increasing function of the Wnt protein concentration v_w, the Wnt protein degradation rate $d_{wp}(v_c)$ is an increasing function of the Cerl2 protein concentration v_c. These functions are sigmoidal functions constructed based on the Hill function.

Assuming as in the previous section that the mRNA dynamics is sufficiently fast and at equilibrium, we obtain a system of differential equations in two variables for (v_w, v_c)

$$\frac{dv_c}{dt} = \frac{a_r c_r}{d_{cr}(S; v_w)} - d_{cp} v_c \tag{3.5a}$$

$$\frac{dv_w}{dt} = \frac{c_w f_w(v_w)}{d_{wr}} - d_{wp}(v_c) v_w. \tag{3.5b}$$

Figure 3.6 shows the nullcline analysis on the (v_w, v_c) plane for the dynamics defined by (3.5). It is drawn using one of the parameter sets that reproduce the observed behavior in the experiment.

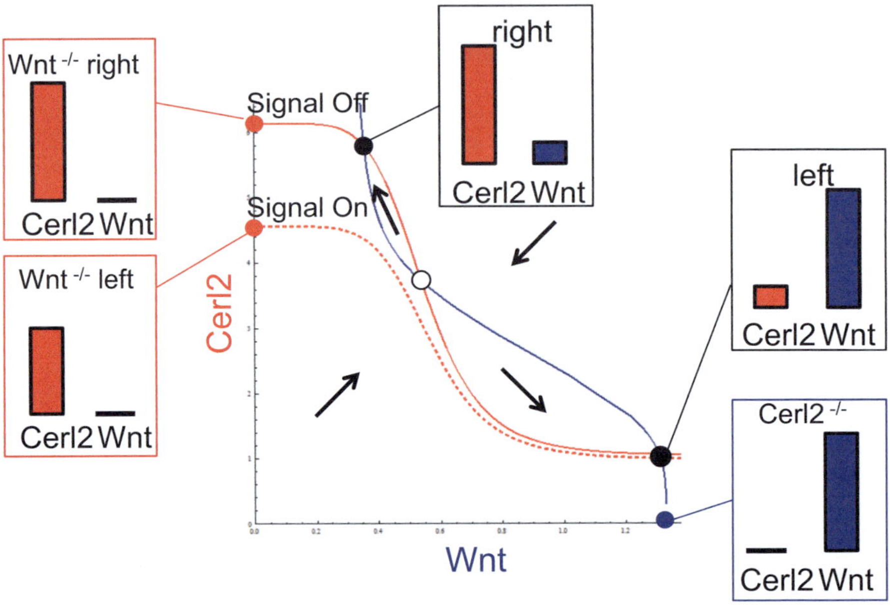

Fig. 3.6 Nullcline analysis for the dynamics (3.5) of mammalian left-right asymmetry. The blue and red curves are nullclines for the Wnt protein (v_w) and Cerl2 protein (v_c), respectively. Cerl2 protein increases at the lower side of the red curve and decreases at the upper side of the red curve, while Wnt protein increases at the lower side of the blue curve and decreases at the upper side of the blue curve. The solid curve indicates before receiving the signal ($S = 0$), and the dashed curve indicates after receiving the signal ($S > 0$). The black circles indicate the stable equilibrium points and the white circles the unstable equilibrium points. The blue point is the equilibrium when the Cerl2 gene is knocked out. The red dots indicate the equilibria at the right and left sides of the Wnt gene knockout

The horizontal axis shows the expression level of Wnt, and the vertical axis shows the expression level of Cerl2. The intersection of the two nullclines (blue curve and red curve) indicates the equilibrium point, especially the black circle, which indicates the stable equilibrium point. An equilibrium point can be interpreted as a state of sustained gene expression. From this figure, we understand that the Wnt-Cerl2 system has two gene expression patterns as stable states in advance before receiving a signal. The expression pattern observed on both the left and right before the signal reception is the upper left state on the plane.

Next, in the cell on the left side receiving the signal, the degradation of Cerl2 mRNA is accelerated, which changes the shape of the Cerl2 nullcline (red curve) from the solid curve to the dashed curve in the figure. At this time, the dashed curve and the blue curve intersect at only one point. This means that the stable equilibrium point (upper left) before receiving the signal is lost. Therefore, the system is forced to move to the other stable expression state (lower right). This is the mechanism by which gene expression changes drastically on the left sides of the node after

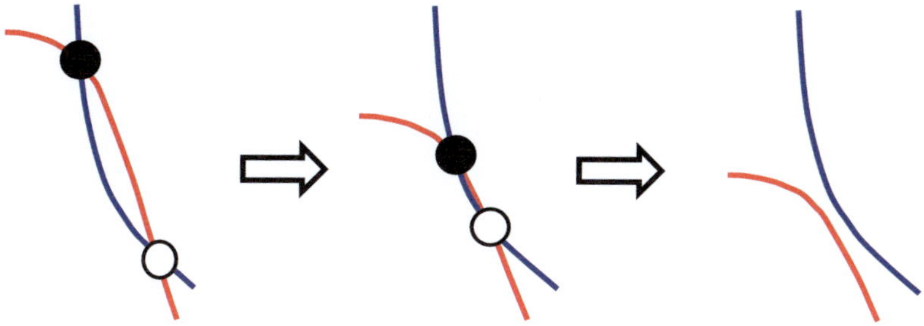

Fig. 3.7 Change in nullclines of Cerl2 and equilibrium points with increasing signal S. Saddle node bifurcation

receiving the signal. Figure 3.6 also illustrates the gene expression of mutants when the Cerl2 gene is knocked out or when the Wnt gene is knocked out.

Let's examine the process of disappearance of the stable equilibrium point more carefully. The nullcline (red curve) of Cerl2 changes continuously as the signal gradually increases. As depicted in Fig. 3.7, the stable equilibrium point and the unstable equilibrium point approach each other and merge and disappear, resulting in a saddle-node bifurcation. In other words, the large change in gene expression seen in the cell on the left side of the node due to the receipt of a signal can be understood as a saddle-node bifurcation.

3.3 Critical Regulation of Left-Right Asymmetric Gene Expression

In this section, we analyze the mathematical model to identify the important regulation for the function of a bistable switch in the Wnt-Cerl2 system. In addition to the production of proteins from mRNAs, the system includes three other regulatory pathways. These are (1) promotion of Cerl2 mRNA degradation by Wnt proteins, (2) promotion of Wnt protein degradation by Cerl2 proteins, and (3) activation of Wnt transcription by Wnt proteins (Fig. 3.5). In particular, (1) is known to be accelerated by nodal flow signals as described above. These regulations constitute two positive feedbacks, both of which appear to be important. Is any one of these three particularly important?

The three regulations are represented in the mathematical model by (1) $d_{cr}(S; v_w)$ being an increasing function of v_w, (2) $d_{wp}(v_c)$ being an increasing function of v_c, and (3) $f_w(v_w)$ being an increasing function of v_w, respectively. By making these functions variable-independent, we can hypothetically consider a system in which each regulation does not work. For example, by making d_{cr} independent of v_w ($d_{cr}(S; v_w) \to d_{cr}(S)$), we can virtually block the promotion of degradation of Cerl2 mRNA by Wnt proteins in (1).

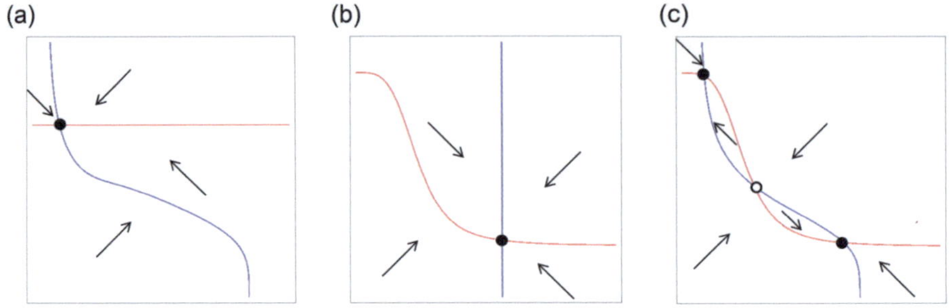

Fig. 3.8 Nullclines for the three models excluding one of the three regulations. (**a**) model without regulation (1), (**b**) model without regulation (2), and (**c**) model without regulation (3). The colors of nullclines indicate the same as in Fig. 3.5

Figure 3.8 displays the nullclines of the model excluding each of the three regulations. Without regulation (1) or (2), one of the two nullclines would be linear, and the system would not be bistable for most parameter values. In this case, a change in the magnitude of the signal S does not cause a significant change in the expression levels of both of the two genes. For (2), depending on the parameter value, the nullcline of v_c may be multiple straight lines. Even in this case, no bifurcation occurs due to the signal change. On the other hand, removing regulation (3) does not significantly change the shape of the nullclines, and large changes in expression due to saddle-node bifurcation can still occur. Therefore, for the Wnt-Cerl2 system to produce large differences in expression levels between the left and right sides, both (1) promotion of Cerl2 mRNA degradation by the Wnt protein and (2) promotion of Wnt protein degradation by the Cerl2 protein are essential. On the other hand, (3) activation of Wnt transcription by Wnt proteins may be less important.

3.4 Experimental Verification

In the analysis shown in Fig. 3.6, many numerical simulations were performed for various parameter values to search for conditions under which the behavior observed in the experiment is realized. The exhaustive search for parameter conditions revealed that there is in fact another possible mechanism, other than bistable switches, that could realize large expression changes depending on the presence or absence of the signal. Even if there is only one stable equilibrium point, the value of the point changes depending on the magnitude of the signal S, because Eq. (3.5a) depends on S. This change can be large under some specific choices of the parameter values. Nakamura et al. call this a “temporal transition to asymmetric state” in the sense of a signal-dependent change of state. Since the parameter conditions under which the temporal transition is realized are very limited in the exhaustive analysis, and a large signal change is required to produce a large change in expression, this

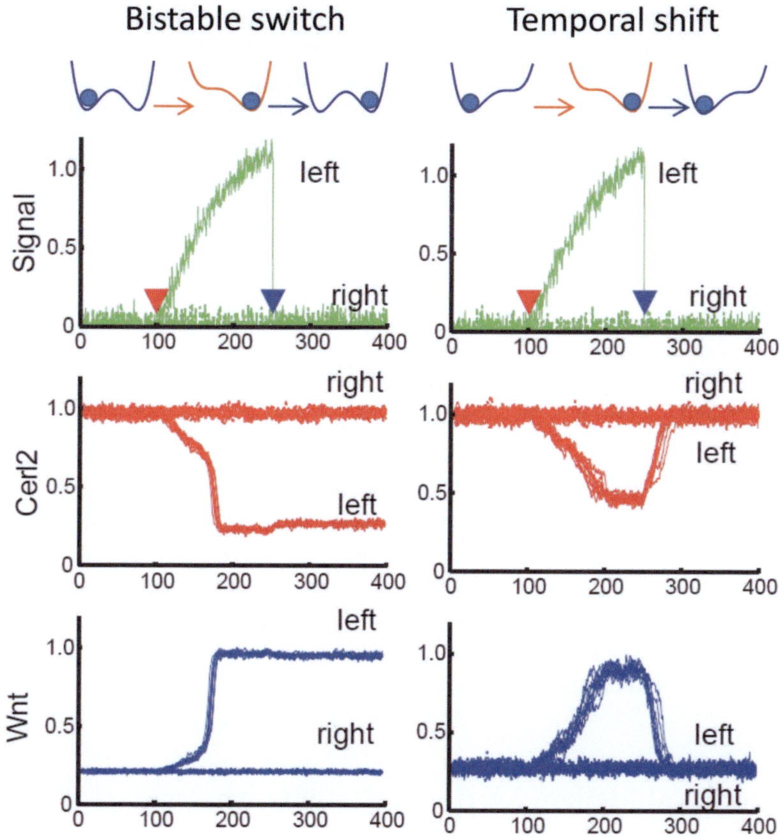

Fig. 3.9 Predicted dynamics of gene expressions in the experiment of forcing the signal to stop. Numerical calculations were performed with noise added. The upper part of the figure shows the changes in the stable states before signal, with the signal, and after the signal termination. (From Nakamura et al. 2012)

mechanism seemed to be less possible in actual mammals. However, the possibility of this mechanism cannot be completely ruled out theoretically. In order to determine whether bistable switches or temporal transitions occur in the real biological system, Nakamura et al. conducted an experiment in which the nodal flow was disrupted in the middle of development. The addition of methylcellulose increases the viscosity of the liquid on the nodal surface and stops the nodal flow (Fig. 3.9).

If this system is realized by bistable switches, then even if the nodal flow is stopped after some time has passed since the signal was received by the left cell, the expression state on the left side will be maintained. This is because both the right and left side expression states are stable equilibrium points. On the other hand, if the left side is realized by the mechanism of temporal transition dependent on the nodal flow signal, the left side system will return to its original right-side state when the nodal flow is terminated.

As a result of this verification experiment, it was found that even if methylcellulose is given during development, the left-right asymmetry of gene expression is the same as that of normal embryos. This confirmed that the asymmetry in gene expression is produced based on the bistable switch in actual mouse embryos (Nakamura et al. 2012). This experiment also clearly demonstrated that the bistable switch mechanism can stably produce large left-right differences even in noisy environments.

3.5 Summary

In this chapter, we introduced the two-variable dynamical system, and as an example of its application to an actual biological system, we discussed left-right asymmetric expression in mice. By using mathematical methods, it is possible to know what kind of dynamics arise from an experimentally obtained gene regulatory system. It is also possible to obtain a deeper understanding of the essential structure of the system for the emergence of the behavior of interest. In the example presented here, we were also able to propose an experiment to compare the validity between two candidate mechanisms. The analysis used to obtain these results is not mathematically advanced. Even simple analyses can yield biologically important information and predictions through creative approaches.

Mathematical Method 3A (Local Stability Analysis)

1. What is local stability analysis?

Consider a system of two-variable ordinary differential equations:

$$\frac{du}{dt} = f(u, v) \tag{3A.1a}$$

$$\frac{dv}{dt} = g(u, v) \tag{3A.1b}$$

where f and g are continuous differentiable nonlinear functions.

In the system of differential Eqs. (3A.1), consider the dynamics near the equilibrium point (u^*, v^*), that is, the point where $f(u^*, v^*) = 0$ and $g(u^*, v^*) = 0$. An equilibrium point is said to be asymptotically stable when the solution approaches the point with time near the equilibrium point. On the other hand, an equilibrium point is said to be unstable when the solution moves away from the point near the equilibrium point. In addition to being stable or unstable, there are also different ways of moving away from or approaching the equilibrium point. The method for

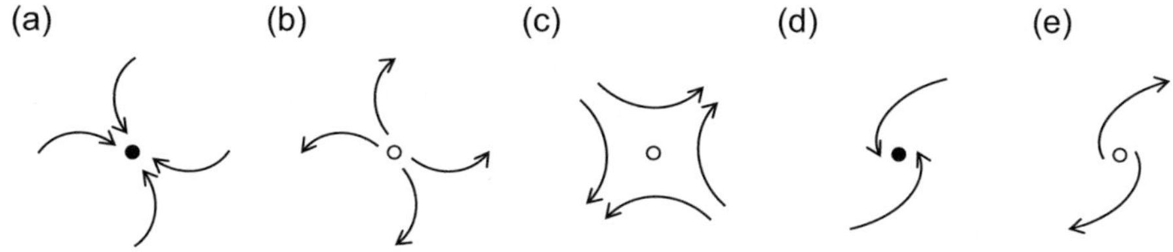

Fig. 3.A1 Five typical behaviors near the equilibrium point in a two-dimensional dynamical system. (**a**) stable node, (**b**) unstable node, (**c**) saddle, (**d**) stable focus, (**e**) unstable focus

determining the behavior of the solution near the equilibrium point is a local stability analysis (Fig. 3.A1) (Murray 2003a, b; Strogatz 2014).

2. Linearization

For the system of differential equations in two variables shown in Eq. (3A.1), consider an approximate system in neighborhood of the equilibrium point (u^*, v^*). Let $(U, V) := (u - u^*, v - v^*)$ be the deviations from the equilibrium point, and substitute into Eq. (3A.1). The left side of Eq. (3A.1) is:

$$\frac{d(U + u^*)}{dt} = \frac{dU}{dt} \quad (3A.2a)$$

$$\frac{d(V + v^*)}{dt} = \frac{dV}{dt} \quad (3A.2b)$$

For the right side of Eq. (3A.1), consider U and V to be minute quantities, and take Taylor expansions for these. Here, the Taylor expansion for two variables is:

$$f(U + u^*, V + v^*) = f(u^*, v^*) + U\frac{\partial f}{\partial u} + V\frac{\partial f}{\partial v} + \cdots \quad (3A.3a)$$

$$g(U + u^*, V + v^*) = g(u^*, v^*) + U\frac{\partial g}{\partial u} + V\frac{\partial g}{\partial v} + \cdots \quad (3A.3b)$$

Where the partial derivatives $\partial f/\partial u$, $\partial f/\partial v$, $\partial g/\partial u$, and $\partial f/\partial v$ are the values at the equilibrium point (u^*, v^*). Also, the first terms on the right side of Eqs. (3A.3a) and (3A.3b) are zero because (u^*, v^*) is in equilibrium. From (3A.2) and (3A.3), Eq. (3A.1) can be approximated in a simple form near the equilibrium point as follows

$$\frac{d}{dt}\begin{pmatrix} U \\ V \end{pmatrix} = \boldsymbol{J}\begin{pmatrix} U \\ V \end{pmatrix} \quad (3A.4a)$$

$$\boldsymbol{J} := \begin{pmatrix} \left.\frac{\partial f}{\partial u}\right|_{(u^*,v^*)} & \left.\frac{\partial f}{\partial v}\right|_{(u^*,v^*)} \\ \left.\frac{\partial g}{\partial u}\right|_{(u^*,v^*)} & \left.\frac{\partial g}{\partial v}\right|_{(u^*,v^*)} \end{pmatrix} \quad (3A.4b)$$

The matrix $\boldsymbol{J}$ is called the Jacobian matrix. In the linear dynamical system (3A.4), the origin $(U, V) = (0, 0)$ is the equilibrium point, which corresponds to the equilibrium point $(u, v) = (u^*, v^*)$ in the original system (3A.1). Even a highly nonlinear system of differential equations can be approximated to a constant coefficient linear ordinary differential equation near the equilibrium point, and its behavior is determined by the Jacobian matrix $\boldsymbol{J}$.

3. Solution of constant coefficient linear ordinary differential equations

Let us consider a system of linear ordinary differential equations

$$\frac{d}{dt}\begin{pmatrix} U \\ V \end{pmatrix} = \boldsymbol{J}\begin{pmatrix} U \\ V \end{pmatrix} := \begin{pmatrix} a & b \\ c & d \end{pmatrix}\begin{pmatrix} U \\ V \end{pmatrix} \tag{3A.5a}$$

$$\begin{pmatrix} U(0) \\ V(0) \end{pmatrix} = \begin{pmatrix} U_0 \\ V_0 \end{pmatrix} \tag{3A.5b}$$

The solution of Eq. (3A.5) is given by

$$\begin{aligned} \begin{pmatrix} U(t) \\ V(t) \end{pmatrix} &= e^{tJ}\begin{pmatrix} U_0 \\ V_0 \end{pmatrix} \\ &:= \left(E + t\boldsymbol{J} + \frac{t^2\boldsymbol{J}^2}{2} + \frac{t^3\boldsymbol{J}^3}{3!} + \cdots\right)\begin{pmatrix} U_0 \\ V_0 \end{pmatrix} = \sum_{k=0}^{\infty} \frac{t^k\boldsymbol{J}^k}{k!}\begin{pmatrix} U_0 \\ V_0 \end{pmatrix} \end{aligned} \tag{3A.6}$$

Where E is the unit matrix. Substitute Eq. (3A.6) and check that Eq. (3A.5) is satisfied (Exercise 3.4). In the following, we assume that the matrix $\boldsymbol{J}$ is diagonalizable; let the two eigenvalues of $\boldsymbol{J}$ be λ_1 and λ_2 and the two corresponding eigenvectors be $(u_1, v_1)^T$ and $(u_2, v_2)^T$, respectively. Then from $\boldsymbol{J}(u_1, v_1)^T = \lambda_1(u_1, v_1)^T$ and $\boldsymbol{J}(u_2, v_2)^T = \lambda_2(u_2, v_2)^T$, we obtain

$$\boldsymbol{J}P = P\begin{pmatrix} \lambda_1 & 0 \\ 0 & \lambda_2 \end{pmatrix} \tag{3A.7a}$$

$$P := \begin{pmatrix} u_1 & u_2 \\ v_1 & v_2 \end{pmatrix} \tag{3A.7b}$$

therefor

$$\boldsymbol{J}^k = P\begin{pmatrix} \lambda_1^k & 0 \\ 0 & \lambda_2^k \end{pmatrix}P^{-1} \tag{3A.8}$$

From (3A.6) and (3A.8), the general solution of Eq. (3A.5) is

$$\begin{pmatrix} U(t) \\ V(t) \end{pmatrix} = P \begin{pmatrix} \sum_{k=0}^{\infty} \frac{t^k \lambda_1^k}{k!} & 0 \\ 0 & \sum_{k=0}^{\infty} \frac{t^k \lambda_2^k}{k!} \end{pmatrix} P^{-1} \begin{pmatrix} U_0 \\ V_0 \end{pmatrix}$$
$$= P \begin{pmatrix} e^{\lambda_1 t} & 0 \\ 0 & e^{\lambda_2 t} \end{pmatrix} P^{-1} \begin{pmatrix} U_0 \\ V_0 \end{pmatrix} \tag{3A.9}$$

Since the two eigenvectors are linearly independent, using two undetermined constants c_1 and c_2, the initial condition can be uniquely expressed as $(U_0, V_0)^T = c_1(u_1, v_1)^T + c_2(u_2, v_2)^T$, that is, $(U_0, V_0)^T = P(c_1, c_2)^T$. This allows the solution to be

$$\begin{pmatrix} U(t) \\ V(t) \end{pmatrix} = P \begin{pmatrix} e^{\lambda_1 t} & 0 \\ 0 & e^{\lambda_2 t} \end{pmatrix} \begin{pmatrix} c_1 \\ c_2 \end{pmatrix} = c_1 e^{\lambda_1 t} \begin{pmatrix} u_1 \\ v_1 \end{pmatrix} + c_2 e^{\lambda_2 t} \begin{pmatrix} u_2 \\ v_2 \end{pmatrix} \tag{3A.10}$$

The form of Eq. (3A.10) is easy to interpret. First, consider the case where the two eigenvalues are real numbers. The solution $(U(t), V(t))^T$ is divided into two components: two distances along directions $(u_1, v_1)^T$ and $(u_2, v_2)^T$, each changing in time according to $e^{\lambda_1 t}$ and $e^{\lambda_2 t}$. The solution can be understood as the sum of these two components. The behavior of the solution can be classified in three ways depending on the positive and negative values of λ_1 and λ_2.

(i) For $\lambda_1 < 0$ and $\lambda_2 < 0$. For each direction of $(u_1, v_1)^T$ and $(u_2, v_2)^T$, the distance from the origin decreases with time according to $e^{\lambda_1 t}$ and $e^{\lambda_2 t}$, and converges to 0 at $t \longrightarrow \infty$. In this case, the origin is asymptotically stable.

(ii) For $\lambda_1 > 0$ and $\lambda_2 > 0$. The distance to the origin increases with $e^{\lambda_1 t}$ and $e^{\lambda_2 t}$ in each direction for$(u_1, v_1)^T$ and $(u_2, v_2)^T$, respectively. If the magnitudes of λ_1 and λ_2 are different, the path bends away from the origin.

(iii) For $\lambda_1 < 0$ and $\lambda_2 > 0$. The distance along $(u_1, v_1)^T$ direction decreases and approaches 0 according to $e^{\lambda_1 t}$. The distance along $(u_2, v_2)^T$ direction increases with $e^{\lambda_2 t}$. As a whole, the solution approaches the origin from the $(u_1, v_1)^T$ direction, but it changes direction near the origin and moves away in the $(u_2, v_2)^T$ direction.

If the eigenvalues λ_1 and λ_2 are complex numbers, we can further transform Eq. (3A.10). Since the matrix $\boldsymbol{J}$ is a real matrix, let $\lambda_1 = \mu + i\omega$ and $\lambda_2 = \mu - i\omega$ be the two eigenvalues that are conjugate complex numbers. Then we obtain

$$\begin{pmatrix} U(t) \\ V(t) \end{pmatrix} = c_1 e^{(\mu+i\omega)t} \begin{pmatrix} u_1 \\ v_1 \end{pmatrix} + c_2 e^{(\mu-i\omega)t} \begin{pmatrix} u_2 \\ v_2 \end{pmatrix}$$
$$= c_1 \begin{pmatrix} u_1 \\ v_1 \end{pmatrix} e^{\mu t}(\cos\omega t + i\sin\omega t) + c_2 \begin{pmatrix} u_2 \\ v_2 \end{pmatrix} e^{\mu t}(\cos\omega t - i\sin\omega t)$$
$$= e^{\mu t}\left[\left\{c_1 \begin{pmatrix} u_1 \\ v_1 \end{pmatrix} + c_2 \begin{pmatrix} u_2 \\ v_2 \end{pmatrix}\right\}\cos\omega t + \left\{c_1 \begin{pmatrix} u_1 \\ v_1 \end{pmatrix} - c_2 \begin{pmatrix} u_2 \\ v_2 \end{pmatrix}\right\} i\sin\omega t\right] \tag{3A.11}$$

where the two components of initial condition $(U_0, V_0)^T = c_1(u_1, v_1)^T + c_2(u_2, v_2)^T$ are both real numbers. Also, since the solution $(U(t), V(t))^T$ is real for $t > 0$, each component of $c_1(u_1, v_1)^T - c_2(u_2, v_2)^T$ is both pure imaginary (Exercise 3.6). Therefore, the solution is

$$\begin{pmatrix} U(t) \\ V(t) \end{pmatrix} = e^{\mu t}\left[\begin{pmatrix} U_0 \\ V_0 \end{pmatrix} \cos \omega t + \begin{pmatrix} W_0 \\ Z_0 \end{pmatrix} \sin \omega t\right] \tag{3A.12}$$

The brackets indicate rotation on a distorted ellipse centered at the origin with an angular velocity ω. The $e^{\mu t}$ in front of the brackets indicates the size of the ellipse. In other words, (3A.12) is a rotating orbit whose amplitude changes with time.

Therefore, when the eigenvalues are complex numbers, there are two possibilities depending on the sign of the real component μ of the eigenvalue.

(iv) For $\mu < 0$. While rotating around the origin, the amplitude gradually decreases and approaches the origin.
(v) For $\mu > 0$. While rotating around the origin, the amplitude gradually increases and moves away from the origin.

4. Local Stability of the Equilibrium Point

The following is a summary of the discussion up to this point. The behavior near the equilibrium point can be approximated to a constant coefficient ordinary differential equation by linearization. By examining the stability of the origin of this equation, we can determine the behavior of the solution near the equilibrium point of the original ordinary differential equation. The stability of the origin (i.e., the equilibrium point) is determined by the eigenvalues of the Jacobian matrix $\boldsymbol{J}$, which is a linearized matrix near the equilibrium point. Therefore, the behavior of the solution near the equilibrium point can be determined by examining the eigenvalues of the Jacobian matrix $\boldsymbol{J}$.

I. The eigenvalues of the Jacobian matrix $\boldsymbol{J}$ are real numbers

(i) Both eigenvalues are negative: The equilibrium point is asymptotically stable, and the solution approaches the equilibrium point with time near the equilibrium point (stable node).
(ii) Both eigenvalues are positive: the equilibrium point is unstable, and the solution moves away from the equilibrium point with time near the equilibrium point (unstable node)
(iii) One eigenvalue is positive and one is negative: The equilibrium point is unstable, and the solution approaches the equilibrium point from one direction and moves away from the equilibrium point in another direction with time (saddle point).

II. The eigenvalues of the Jacobian matrix $\boldsymbol{J}$ are complex

(iv) Real part of eigenvalues is negative: the equilibrium point is asymptotically stable, and the solution approaches the equilibrium point while rotating around the equilibrium point in the neighborhood of the equilibrium point (stable focus).

(v) Real part of eigenvalues is positive: The equilibrium point is unstable, and the solution moves away from the equilibrium point near the equilibrium point while rotating around the equilibrium point (unstable focus).

The eigenvalue λ is given as the solution of the characteristic equation $\lambda^2 - \mathrm{tr}\boldsymbol{J}\lambda + \det\boldsymbol{J} = 0$, using the trace $\mathrm{tr}\boldsymbol{J} = \partial f/\partial u + \partial g/\partial v$ and the determinant $\det\boldsymbol{J} = \partial f/\partial u \cdot \partial g/\partial v - \partial f/\partial v \cdot \partial g/\partial u$ of the Jacobian matrix $\boldsymbol{J}$. Therefore, from the relationship between the solution and the coefficients of the quadratic equation, the condition of stability can also be expressed as follows.

The condition that the equilibrium point is a stable node, i.e., both eigenvalues are real and negative, is given by

$$(\mathrm{tr}\boldsymbol{J})^2 - 4\det\boldsymbol{J} \geq 0 \tag{3A.13a}$$

$$\mathrm{tr}\boldsymbol{J} < 0 \tag{3A.13b}$$

$$\det\boldsymbol{J} > 0 \tag{3A.13c}$$

The condition that the equilibrium point is a stable focus, i.e., the eigenvalues are complex and their real part is negative, is given by

$$(\mathrm{tr}\boldsymbol{J})^2 - 4\det\boldsymbol{J} < 0 \tag{3A.14a}$$

$$\mathrm{tr}\boldsymbol{J} < 0 \tag{3A.14b}$$

Combining the two, the necessary and sufficient conditions for the equilibrium point to be asymptotically stable in a two-dimensional dynamical system are as follows

$$\mathrm{tr}\boldsymbol{J} < 0 \tag{3A.15a}$$

$$\det\boldsymbol{J} > 0 \tag{3A.15b}$$

Figure 3.A2 shows the classification of the behaviors near the equilibrium point by the trace tr$\boldsymbol{J}$ and the determinant det$\boldsymbol{J}$. Local stability analysis is a basic method of dynamical system analysis, but it is also a method that is used repeatedly in various aspects of the analysis.

Mathematical Method 3B (Non-dimensionalization)

After setting up a differential equation model, it is usual to check its behavior by performing numerical calculations for various parameter values and examining the parameter dependence. If the model contains many parameters, analyzing parameter

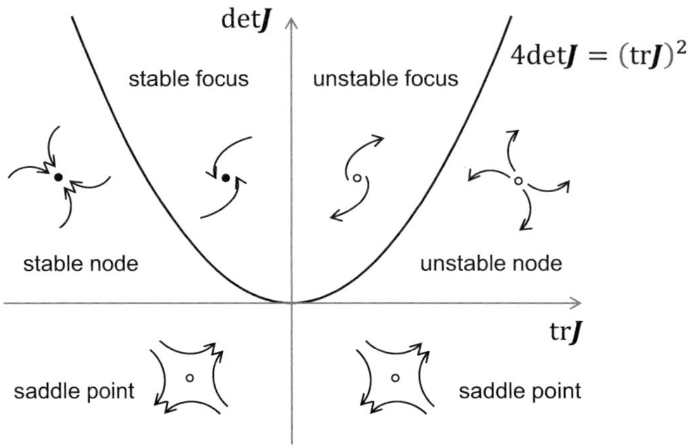

Fig. 3.A2 Classification of the behavior near an equilibrium point of a two-variable dynamical system by trace tr***J*** and determinant det***J*** of the Jacobian matrix

dependence is a difficult task. One way to reduce the degrees of freedom is to reassign dimensional parameters to the variables in the differential equations. Let us take as an example the system of differential equations corresponding to Eq. (3.2) in this chapter:

$$\frac{dv_1}{dt} = \frac{a_1 c_1 / b_1}{1 + (v_2/h_2)^n} - d_1 v_1 \tag{3B.1a}$$

$$\frac{dv_2}{dt} = \frac{a_2 c_2 / b_2}{1 + (v_1/h_1)^n} - d_2 v_2 \tag{3B.1b}$$

Perform variable transformation for each of the state variables (u, v) and time t:

$$\begin{aligned} v_1 &= \alpha v_1' \\ v_2 &= \beta v_2' \\ t &= \gamma t' \end{aligned} \tag{3B.2}$$

At this point, assume that α, β, and γ are arbitrary constants. The dimensions (mol/l, second, μm, etc.) of v_1, v_2, and t are included in α, β, and γ, and v_1, v_2, and t are assumed to be dimensionless quantities. This is why this method is called dimensionless. Substituting (3B.2) into (3B.1) and rearranging, we obtain

$$\frac{dv_1'}{dt'} = \frac{a_1 c_1 \gamma}{b_1 \alpha} \frac{1}{1 + \left(\beta v_2'/h_2\right)^n} - d_1 \gamma v_1' \tag{3B.3a}$$

$$\frac{dv_2'}{dt'} = \frac{a_2 c_2 \gamma}{b_2 \beta} \frac{1}{1 + \left(\alpha v_1' / h_1\right)^n} - d_2 \gamma v_2' \tag{3B.3b}$$

Let $\alpha / h_1 = 1$, $\beta / h_2 = 1$, and $d_1 \gamma = 1$, i.e.:

$$\begin{aligned} \alpha &= h_1 \\ \beta &= h_2 \\ \gamma &= \frac{1}{d_1} \end{aligned} \tag{3B.4}$$

then Eq. (3B.3) follows:

$$\frac{dv_1'}{dt'} = \frac{A_1}{1 + \left(v_2'\right)^n} - v_1' \tag{3B.5a}$$

$$\frac{dv_2'}{dt'} = \frac{A_2}{1 + \left(v_1'\right)^n} - D v_2' \tag{3B.5b}$$

Where A_1, A_2, and D are newly set parameters and are defined as follows:

$$A_1 := \frac{a_1 c_1 \gamma}{b_1 \alpha} = \frac{a_1 c_1}{b_1 d_1 h_1} \tag{3B.6a}$$

$$A_2 := \frac{a_2 c_2 \gamma}{b_2 \beta} = \frac{a_2 c_2}{b_2 d_1 h_2} \tag{3B.6b}$$

$$D := d_2 \gamma = \frac{d_2}{d_1} \tag{3B.6c}$$

The original Eq. (3B.1) contains seven parameters, whereas the non-dimensionalized Eq. (3B.5) contains only four parameters. The parameter dependence of the dynamics based on this equation should be exhausted by examining these four parameters. To understand the biological meaning of A_1, A_2, and D, we can go back to the definition of parameters (3B.6).

Let us summarize the operations. By assuming that α, β, and γ, originally set as arbitrary constants, were actually the quantities as in (3B.4), we have reduced the number of parameters in the original dynamical system. This means that the dimension is reduced from the apparent number of parameters when the coefficients of the variable transformation are special values. Since the objective here is to reduce the number of parameters, we can choose the coefficients we wish to eliminate in step (3B.3). The number of parameters can be reduced up to the number of transformed variables.

The reduction of parameter dimension by non-dimensionalization can be intuitively understood as follows. The effects of multiple parameters in the original dynamical system are not independent of each other. For example, simultaneously doubling the parameters a_1, a_2, d_1, and d_2 on the right-hand side of the equation is equivalent to doubling the rate of progress of time. Reducing the dimension of parameters by scaling variables is equivalent to eliminating redundant examination for behaviors that are identifiable by the scaling transformation.

Exercise

Exercise 3.1

For the system of four-variable differential Eqs. (3.1), set $du_1/dt = 0$ and $du_2/dt = 0$ to obtain the system of two-variable differential Eqs. (3.2). Then, by applying nondimensionalization, obtain the system of differential Eqs. (3.3).

Exercise 3.2

Let $(\widehat{v}_1, \widehat{v}_2)$ be a hypothetical equilibrium point of the differential equation system (3.3), and find the linearized equation near the point. Show that the condition for the equilibrium point $(\widehat{v}_1, \widehat{v}_2)$ to be stable is given by $n^2 A_1 A_2 \widehat{v}_1^{n-1} \widehat{v}_2^{n-1} < D\left(1 + \widehat{v}_1^n\right)^2 \left(1 + \widehat{v}_2^n\right)^2$.

Exercise 3.3

For the system of differential equations in (3.5), (1) show that the nullcline of v_c becomes a straight line as in Fig. 3.8a by making $d_{cr}(S; v_w)$ a function that does not depend on v_w. Similarly, (2) show that the nullcline of v_w becomes (a) straight line (s) as shown in Fig. 3.8b by making $d_{wp}(v_c)$ a function that does not depend on v_c. (Depending on the parameter values, the nullcline of v_w may be multiple lines. In that case, there is no bifurcation due to the presence or absence of the signal.) On the other hand, even if (3) $f_w(v_w)$ is made a function that does not depend on v_w, show that the nullcline of v_c and the nullcline of v_w are not always straight lines.

Exercise 3.4

Confirm that Eq. (3A.6) is a solution to the linear ordinary differential equation system (3A.5). Substitute into each side of (3A.5a) and show that both sides are equal, and that (3A.5b) is satisfied at $t = 0$.

Exercise 3.5

Derive the form of Eq. (3A.10) from the form of Eq. (3A.6) which is the solution of the linear ordinary differential equation system (3A.5).

Exercise 3.6

Consider the case where the Jacobian matrix $\boldsymbol{J}$ of the linear ordinary differential Eq. (3A.5) has complex eigenvalues. Let λ_1 and λ_2 be the conjugate complex eigenvalues of the matrix $\boldsymbol{J}$. Show that the two corresponding eigenvectors $(u_1, v_1)^T$ and $(u_2, v_2)^T$ are such that u_1 and u_2, and v_1 and v_2 are conjugate complex numbers, respectively. From this, since the initial conditions $c_1(u_1, v_1)^T + c_2(u_2, v_2)^T$ are real numbers, $c_1 = c_2$ is required. Using the above, derive the form of (3A.11) to (3A.12).

Exercise 3.7

Show that the condition (3A.15) for the stability of the equilibrium point of a two-variable dynamical system is derived from the condition (3A.13) for the case where the eigenvalues are real and the condition (3A.14) for the case where the eigenvalues are complex. Conversely, show that the condition (3A.13) or the condition (3A.14) is derived from the condition (3A.15).

References

Arkin A, Ross J, McAdams HH (1998) Stochastic kinetic analysis of developmental pathway bifurcation in phage lambda-infected Escherichia coli cells. Genetics 149:1633–1648

Murray JD (2003a) Mathematical biology, vol I, 3rd edn. Springer, New York, p 574

Murray JD (2003b) Mathematical biology, vol II, 3rd edn. Springer, New York, p 839

Nakamura T, Saito D, Kawasumi A, Shinohara K, Asai Y, Takaoka K, Dong F, Takamatsu A, Belo JA, Mochizuki A, Hamada H (2012) Fluid flow and interlinked feedback loops establish left-right asymmetric decay of Cerl2 mRNA in the mouse embryo. Nat Commun 3:1322

Nonaka S, Shiratori H, Saijoh Y, Hamada H (2002) Determination of left-right patterning of the mouse embryo by artificial nodal flow. Nature 418:96–99

Ptashne M (2004) Genetic switch: phage lambda revisited. Cold Spring Harbor Laboratory Press, Cold Spring Harbor, p 154

Strogatz S (2014) Nonlinear dynamics and chaos, 2nd edn. Westview Press, p 532

Chapter 4
Oscillation in Living System I: Circadian Rhythms and Segmentation Clocks

Abstract There are several examples of biological phenomena in which biological functions arise from periodic oscillations of biomolecular activities over time. Here, we will discuss well-known oscillatory phenomena such as circadian rhythms and segmentation clocks and their molecular mechanisms to clarify how biological functions are generated from the viewpoint of dynamics.

Keywords Circadian rhythm · Negative feedback · Hopf bifurcation · Limit cycle · Global stability · Segmentation clock · Time delay

4.1 Circadian Rhythms and Negative Feedback

The circadian rhythm is a roughly 24-hour oscillation seen in the physiological functions of living organisms. We feel jet lag when we travel abroad because there is a time difference between the external environment and the clock that ticks time autonomously within our bodies. In the last 30 years or so, the mechanisms of circadian rhythms have been clarified at the molecular level in many species of living organisms (Dunlap 1999; Young and Kay 2001; Hardin et al. 1990). In this process, mathematical models have also played an important role. A typical example of mathematical research is the series of work by Goldbeter and colleagues. It is now known that the molecular mechanisms of circadian rhythms in many species can be understood on a common principle. The oscillations are understood to be caused by the negative feedback regulation of gene transcription.

Figure 4.1 shows the basic framework of gene regulation of circadian rhythms in Drosophila. Genes that play a major role in circadian rhythms are called clock genes. In Drosophila, two clock genes, *per* and *tim*, are transcribed and translated into proteins outside the nucleus. After modification, these proteins form a heterodimer and are transported into the nucleus to repress transcription of both *per* and *tim*. This regulation, in which transcription is repressed by its own product, is called "negative feedback."

The following is an intuitive explanation of the oscillations caused by the negative feedback. For simplicity, let us group the two genes together as if they were one and focus only on transcription rate and protein concentration. (i) First,

A. Mochizuki, *Structural Theories for Dynamical Networks in Biology*, Theoretical Biology, https://doi.org/10.1007/978-981-96-6026-1_4

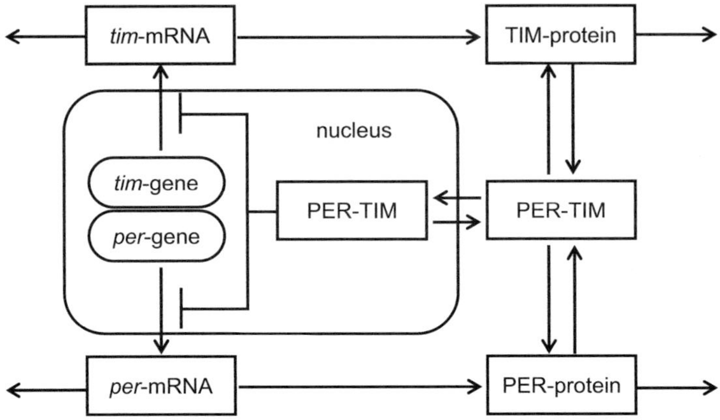

Fig. 4.1 Regulation of the clock genes that underlies circadian rhythms in Drosophila. Transcription factors, products of genes, repress their own transcription. In Drosophila genetics, genes are written in lowercase italics and proteins in uppercase solid letters. (Based on Kurosawa et al. 2002)

let's consider a state where transcription is actively taking place. Then, the protein concentration gradually increases. (ii) As the protein concentration increases, the negative feedback inhibition of transcription also becomes stronger. Therefore, the transcription rate gradually decreases. (iii) As the transcription level decreases, the protein synthesis rate also decreases, and the protein concentration decreases gradually due to degradation. (iv) As the protein concentration decreases, the transcriptional repression is released, and transcription becomes active again. This cycle of events results in periodic gene activity.

Figure 4.2 shows an example of a periodic oscillation obtained by numerical simulation of a mathematical model for the negative feedback regulation of the per gene in Drosophila.

The conditions for the existence of periodic oscillation can be determined to some extent by mathematical analysis, although it depends on the choice of parameter values. Let us check the conditions under which the periodic oscillation solution appears using another mathematical model. Kurosawa et al. constructed and analyzed a differential equation model for the time changes of *per* and *tim* mRNAs, proteins, and dimers according to the regulation shown in Fig. 4.1.

$$\frac{dM_1}{dt} = \frac{1}{1+(P/h)^n} - a_1 M_1 \tag{4.1a}$$

$$\frac{dM_2}{dt} = \frac{1}{1+(P/h)^n} - a_2 M_2 \tag{4.1b}$$

$$\frac{dR_1}{dt} = sM_1 - d_1 R_1 - bR_1 R_2 + cP \tag{4.1c}$$

$$\frac{dR_2}{dt} = sM_2 - d_2 R_2 - bR_1 R_2 + cP \tag{4.1d}$$

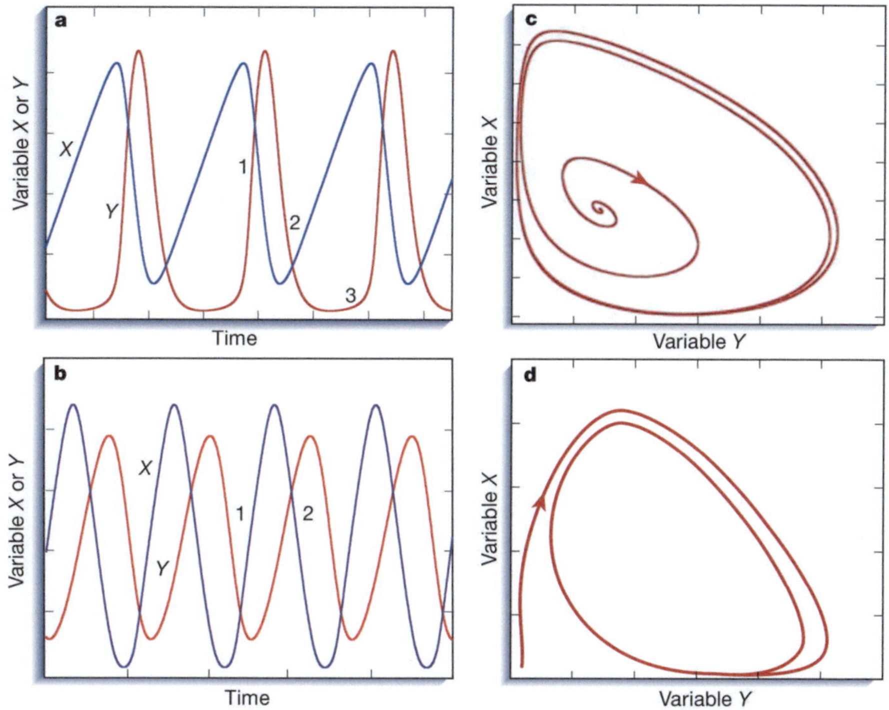

Fig. 4.2 An example of periodic oscillations based on a negative feedback model for circadian rhythms. (Goldbeter 2002)

$$\frac{dP}{dt} = bR_1R_2 - cP \tag{4.1e}$$

where M_1 and M_2 are mRNA concentrations of *per* and *tim*, respectively; R_1 and R_2 are concentrations of PER and TIM proteins, respectively; and P is the concentration of TIM-PER protein heterodimer. a_1, a_2, d_1, and d_2 are mRNA and protein degradation rate coefficients, s is the translation rate coefficients, and b and c are the rate coefficients of the reversible complex formation reaction in the forward and backward directions, respectively. The first term (transcription) in Eqs. (4.1a) and (4.1b) is the form of 1 minus the Hill function, which we learned in Chap. 2, and it rapidly decreases and approaches 0 when P exceeds the threshold value h.

Let us assume that there is no difference in the rate constant of the reactions between the two genes, *per* and *tim*. In this case, after a sufficient time, we can consider a three-dimensional dynamical system as if it were the dynamics of a single gene (Exercises 4.1 and 4.2):

$$\frac{dM}{dt} = \frac{1}{1 + (P/h)^n} - aM \tag{4.2a}$$

$$\frac{dR}{dt} = sM - dR - bR^2 + cP \tag{4.2b}$$

$$\frac{dP}{dt} = bR^2 - cP \tag{4.2c}$$

where M is the mRNA concentration, R is the protein concentration, and P is the dimer concentration. The parameters a, b, c, d, and s are all assumed to be positive. There exists an equilibrium point $\left(\widehat{M}, \widehat{R}, \widehat{P}\right)$ with positive values in this dynamical system. By placing $m{:=}M - \widehat{M}$, $r{:=}R - \widehat{R}$, and $p{:=}P - \widehat{P}$, the linearized equations around this equilibrium point are given by:

$$\frac{d}{dt}\begin{pmatrix} m \\ r \\ p \end{pmatrix} = \begin{pmatrix} -a & 0 & \dfrac{-n\left(\widehat{P}/h\right)^{n-1}}{h\left\{1+\left(\widehat{P}/h\right)^n\right\}^2} \\ s & -d-2b\widehat{R} & c \\ 0 & 2b\widehat{R} & -c \end{pmatrix}\begin{pmatrix} m \\ r \\ p \end{pmatrix} \tag{4.3}$$

If this equilibrium point is stable, there is no periodic oscillation in the solution of the system. For periodic oscillations to exist, this equilibrium point must be unstable. The degree of nonlinearity that destabilizes the equilibrium point can be determined analytically. Specifically, for the model in Eq. (4.3), the equilibrium point is always stable for $n \leq 4$, regardless of other parameter values. We see that $n \geq 5$ is required for an oscillatory solution to appear (Exercise 4.4). In other words, a high degree of nonlinearity in the transcription inhibition is required for periodic oscillations. To learn more about the process of instability of the equilibrium point, let us consider a simplified linearized equation:

$$\frac{d}{dt}\begin{pmatrix} m \\ r \\ p \end{pmatrix} = \begin{pmatrix} -d & 0 & -c \\ a & -d & 0 \\ 0 & b & -d \end{pmatrix}\begin{pmatrix} m \\ r \\ p \end{pmatrix} \tag{4.4}$$

Here, the state transition from P to R is ignored, and the decay of M, R, and P are all assumed to have equal rate coefficients. The reaction rate coefficients for the reactions that form the negative feedback $M \to R \to P \to M$ are newly defined as a, b, and c. a, b, c, and d are all positive parameters. The equilibrium point of this linearized equation becomes unstable under the condition $2d < \sqrt[3]{abc}$ (Exercise 4.5). At $2d = \sqrt[3]{abc}$, the equilibrium point changes from stable focus to unstable focus (see 3A Mathematical Methods, Figs. 3.A1 and 3.A2). This transition is called a Hopf bifurcation. Intuitively, a Hopf bifurcation occurs when the negative feedback has a larger effect than the damping term (Fig. 4.3).

In the case of a two-dimensional system, when instability occurs due to Hopf bifurcation, the Poincaré-Bendixon theorem (4B mathematical method) may be used to prove mathematically that stable periodic orbits exist around the unstable

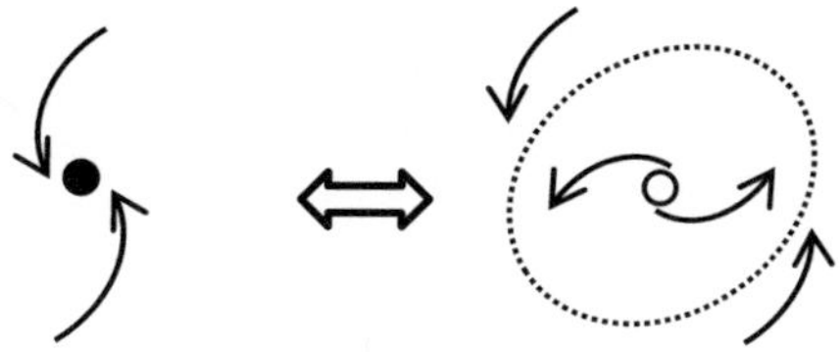

Fig. 4.3 Hopf bifurcation. Transition from stable focus to unstable focus or vice versa

equilibrium point. In the case of the three-variable system (4.2), the existence of periodic oscillations must be verified numerically under conditions where the equilibrium point is unstable.

4.2 Global Stability of the Two-Variable Negative Feedback Model

Negative feedback, in which the transcription rate is suppressed by the product protein, can cause temporal oscillations. Whereas a negative feedback does not always induce temporal oscillations. In this regard, one interesting result (Kurosawa et al. 2002) considers a single-gene self-repression circuit that contains only two variables, i.e., mRNA and protein concentrations.

The dynamics of a system such as shown in Fig. 4.4 can generally be described by a system of differential equations of the form:

$$\frac{dM}{dt} = f(P) - h(M) \tag{4.5a}$$

$$\frac{dP}{dt} = g(M) - k(P) \tag{4.5b}$$

where M and P are the mRNA and protein concentrations, respectively. The transcription function $f(P)$ is a monotonic decreasing function of P. The translation function $g(M)$ and the decomposition terms $h(M)$ and $k(P)$ are monotonic increasing functions of M, M, and P, respectively. The differential equation system (4.5) is assumed to have only one equilibrium point $(M,P) = \left(\widehat{M}, \widehat{P}\right)$ satisfying $f\left(\widehat{P}\right) - h\left(\widehat{M}\right) = 0$, $g\left(\widehat{M}\right) - k\left(\widehat{P}\right) = 0$. Now consider a function on the (M,P) plane:

$$V(M,P) = -\int_{\widehat{P}}^{P} \left(f(x) - f\left(\widehat{P}\right)\right)dx + \int_{\widehat{M}}^{M} \left(g(x) - g\left(\widehat{M}\right)\right)dx \tag{4.6}$$

We see that V is a smooth function with respect to (M,P), (i) it takes a minimum at $\left(\widehat{M}, \widehat{P}\right)$ and (ii) the time derivative dV/dt of V is always nonpositive, and zero only at $(M,P) = \left(\widehat{M}, \widehat{P}\right)$ (Exercises 4.7 and 4.8). In other words, the dynamics of (M,P) goes in the direction of decreasing V and finally approaches the point $\left(\widehat{M}, \widehat{P}\right)$ where

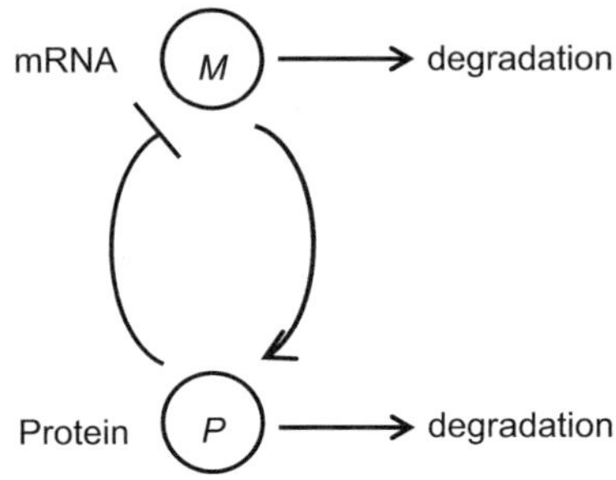

Fig. 4.4 Genetic circuit for a self-repressing gene. M is the mRNA concentration, and P is the protein concentration

V is at its minimum. In other words, the equilibrium point $\left(\widehat{M}, \widehat{P}\right)$ is stable in the sense that the solution approaches not only from its neighborhood but also from the entire space (M, P). This is called globally asymptotically stable. A function satisfying properties such as (i) and (ii) is called a Lyapunov function. Since the equilibrium point $\left(\widehat{M}, \widehat{P}\right)$ is globally stable, we can prove that there is no oscillatory solution for this system. Note that in this discussion, no specific form of the functions $f(P)$, $g(M)$, $h(M)$, and $k(P)$ is assumed, but only monotonicity. In other words, in a regulatory system such as the one shown in Fig. 4.4, no stable periodic oscillations appear, whatever the form of the functions; whatever the initial conditions for the mRNA and protein, the system will converge to an equilibrium point $\left(\widehat{M}, \widehat{P}\right)$ after a sufficient time. To come to this conclusion, it should be noted that the effect of the protein on transcription is only repression ($f(P)$ is a monotonically decreasing with P). In models that include positive regulation on transcription (the transcription function $f(P)$ may increase depending on the value of P), oscillatory solutions may appear.

As shown in Sect. 4.2, periodic oscillations can appear in models that include dimers or multiple genes in addition to mRNA and protein. Periodic oscillations can also appear in a three-variable model that takes into account phosphorylation modification of proteins. Kurosawa et al. classified the conditions for various models to exhibit oscillatory solutions based on the Hill coefficient of the Hill function of the transcription function (Kurosawa et al. 2002). The number of variables, or the dimension of a dynamical system, is one of the important factors that determine the possible behaviors of a system. Some dynamical behaviors may require a certain minimum number of dimensions to be realized. For example, a continuous dynamical system without time delay must contain at least three variables for chaos to appear.

4.3 Segmentation Clock

Vertebrates have periodic structures along their body axis from head to tail, such as vertebrae, ribs, and skeletal muscles. These repeating structures are derived from a series of mesodermal cell blocks of approximately constant size, called segments. Segments are formed during development by separating one by one from a

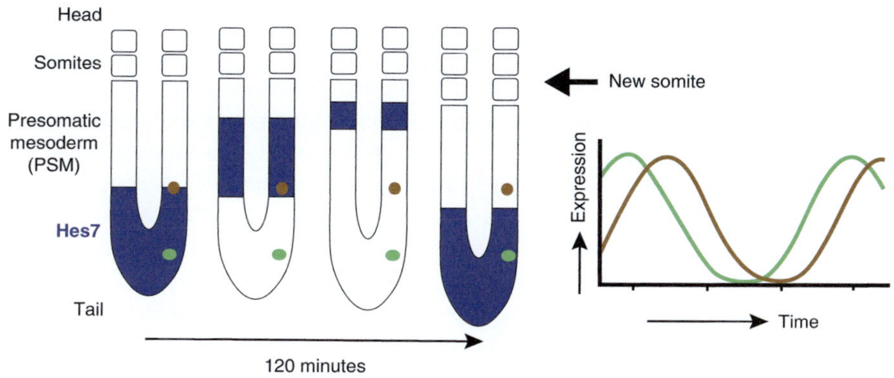

Fig. 4.5 Spatiotemporal dynamics of a segmentation clock. (Courtesy of Ryoichiro Kageyama)

continuous tissue called pre-somitic mesoderm in the head-to-tail direction (Fig. 4.5). This segmentation process is repeated at regular intervals that are unique to each species. The mechanism that controls this periodicity is called the segmentation clock.

The group of Professor Ryoichiro Kageyama (Kyoto University, RIKEN Brain Science Center at present) has analyzed the molecular mechanism of the segmentation clock. They have revealed the importance of the dynamics of a gene called Hes7 and the gene regulatory mechanism that produces the dynamics (Bessho et al. 2001a, b). In mouse pre-somatic mesoderm cells, the expression of Hes7 oscillates in a 120-minute cycle. The timing (phase) of Hes7 expression in the tail is early and that in the head is late. This phase shift is observed as if the expression wave moves from the tail side to the head side (Fig. 4.5). The region of Hes7 expression at the timing when the expression wave reaches its most headward position determines the width of one segment in the later stages. This process creates a spatially periodic structure in which segments are aligned at an almost constant size. In other words, the dynamics of Hes7 is a process in which temporal oscillations are converted into spatially periodic patterns.

Since knockout of Hes7 abolishes temporal oscillations and at the same time causes fusion of segments, Hes7 is thought to be the essence of the oscillations in gene expression and that of the periodic patterns. In addition, the expression of Hes7 was found to be mediated by a negative feedback in which the Hes7 protein represses its own transcription. This was thought to be the fundamental mechanism of the oscillatory expression of Hes7. Based on this idea, they developed a mathematical model (Lewis 2003; Hirata et al. 2004):

$$\frac{dm(t)}{dt} = f(p(t - T_m)) - cm(t) \tag{4.7a}$$

$$\frac{dp(t)}{dt} = am\left(t - T_p\right) - bp(t) \tag{4.7b}$$

$$f(p) = \frac{k}{1 + (p/p_{crit})^2} \tag{4.7c}$$

where $m(t)$ and $p(t)$ are the mRNA and protein concentrations at time t, respectively, and $f(p)$ is the transcription function. Transcription and translation depend on the protein concentration $p(t - T_m)$ at time $t - T_m$ and the mRNA concentration $m(t - T_p)$ at time $t - T_p$, respectively. This type of system is called dynamics with time delay; T_m and T_p are parameters, which are called time delay constants.

If T_m and T_p are zero, the system (4.7) will be of the form (4.5) discussed in the previous section, and the equilibrium point will be globally stable, so periodic oscillations will never appear. On the other hand, it is known analytically that oscillatory solutions can appear when the time delay constants T_m and T_p are sufficiently large. Relevant analytical examples are presented in the 4C mathematical method. In general, it is known that the dynamics of differential equation models can significantly change when time delays are included.

Takashima et al. in the Kageyama group experimentally demonstrated the importance of the time delay using actual mouse embryos (Takashima et al. 2011). The Hes7 gene contains three introns and undergoes intron excision (splicing) during maturation of mRNA after transcription. This splicing process should take a certain duration, and this process may be the molecular basis for the time delay T_m in the mathematical model. Takashima et al. created mutant mice in which the wild-type Hes7 gene was replaced by a gene with the intron predeleted and examined the effect of this mutation. Intron deletion directly shortened the time from transcription initiation to expression by approximately 20 minutes.

In the mutant in which all three introns were deleted, the oscillatory expression of Hes7 was lost, and fusion of all segments was observed. In other words, in order for oscillatory expression to occur, it is necessary for the mRNA synthesis process to take a sufficient time. Harima et al. in the Kageyama group also created a mutant in which only two out of three introns were deleted (Harima et al. 2013). Interestingly, the expression oscillations were not completely lost, and the period of oscillation was shorter than in normal embryos. In this mutant, the time required for segmentation was shortened, so more somites were formed than in normal individuals, and the number of vertebrae also increased. In principle, mammals have seven vertebrae in their necks. However, mice with two introns deleted have nine cervical vertebrae. This result indicates that segmentation clocks caused by oscillation in Hes7 expression play a major role even in the basic body plan. The time delay constant T_m was not just a parameter in a mathematical model, but a directly manipulatable value that represented the "time required for RNA splicing." This fact made it possible to compare numerical simulations with experimental results.

In addition to somite genesis, oscillations in Hes7 gene expression have been shown to play an important role in another aspect of development, neuronal differentiation. In the process of neuronal differentiation, Hes7 regulates the activity of the Dll gene, which in turn regulates the activity of Hes genes in neighboring cells through Delta-Notch signaling. Interestingly, Dll expression exhibits different

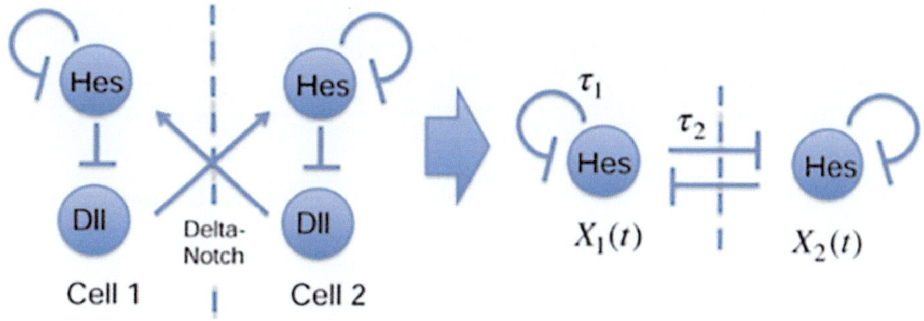

Fig. 4.6 Coupled oscillator model of Hes7. (From Shimojo et al. 2016)

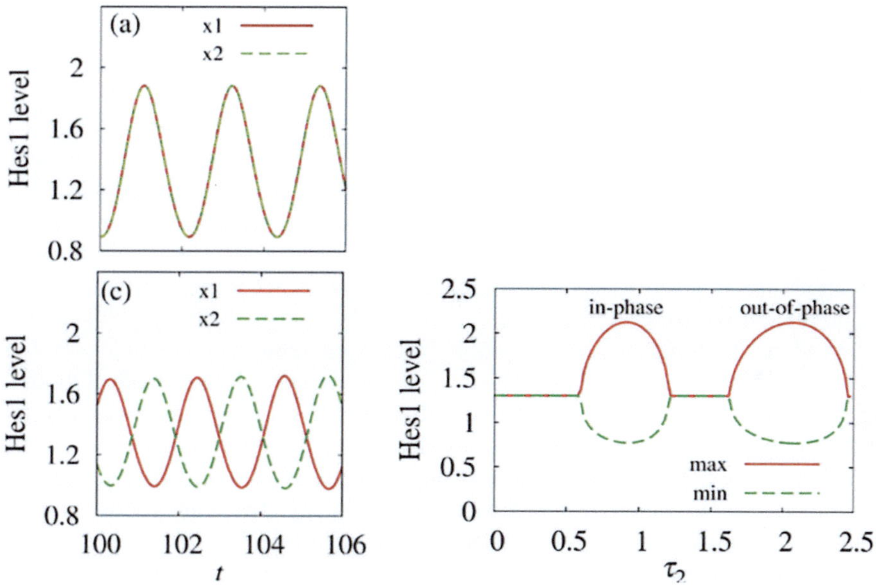

Fig. 4.7 Coupled oscillator model of Hes7. (From Shimojo et al. 2016). (**a**) In-phase oscillation. (**b**) Anti-phase oscillation. (**c**) The vertical and horizontal axes show the maximum and minimum values of u_i and the time delay constant τ_2, respectively. In this figure, in-phase oscillations are observed around $\tau_2 = 1$, and anti-phase oscillations are observed around $\tau_2 = 2$

spatiotemporal dynamics between somite genesis and neuronal differentiation. In somite genesis, Dll expression exhibits synchronous oscillations in a wide region of pre-somitic mesoderm, whereas in neural progenitor cells, it shows desynchronous oscillations in which neighboring cells are out of phase with each other (Fig. 4.6).

Shimojo et al. of the Kageyama group proposed a mathematical model in which Hes oscillations are coupled between two neighboring cells and their activities are mutually inhibited (Shimojo et al. 2016) (Fig. 4.7):

$$\frac{d}{dt}u_1(t) = v\frac{K_1^m}{K_1^m + u_1(t-\tau_1)^m}\frac{K_2^n}{K_2^n + u_2(t-\tau_2)^n} - ru_1(t) \tag{4.8a}$$

$$\frac{d}{dt}u_2(t) = v\frac{K_1^m}{K_1^m + u_2(t-\tau_1)^m}\frac{K_2^n}{K_2^n + u_1(t-\tau_2)^n} - ru_2(t) \tag{4.8b}$$

The model (4.8) includes two time delay constants: τ_1 is the time delay for transcription and translation of Hes7, and τ_2 is the time delay for cell-to-cell inhibitory regulation. Shimojo et al. showed that the relative magnitude of these two time delays causes in-phase or out-of-phase oscillations. These results suggest that pre-somatic mesoderm cells and neuronal differentiation undergo different oscillations due to differences in the time required for intercellular inhibitory regulation.

4.4 Summary

In this chapter, we discussed two important examples of temporal oscillations producing biological functions: circadian rhythms and somite genesis. The basic principle of both mechanisms is negative feedback in the regulatory systems. Some mathematical conditions for the emergence of oscillatory solutions are also presented, and it is shown that the presence of negative feedback does not always lead to the emergence of oscillations.

Circadian rhythms and segmentation clocks are representative fields of life science that have been greatly advanced by the integration of mathematical modeling. In these fields, the use of mathematical models has not only given us a better understanding of the phenomena but has also led to the development and implementation of new experimental ideas from the analysis of mathematical models, which has further advanced our understanding of biology. These are good examples of successful collaboration between theory and experiment.

Mathematical Method 4A (Asymptotic Behavior of Dynamical Systems)

We introduce some concepts on the behavior of dynamical systems after a sufficient time (asymptotic behavior). Let us consider an initial value problem for an N-dimensional dynamical system.

$$\frac{d\boldsymbol{u}}{dt} = F(\boldsymbol{u}) \tag{4A.1a}$$

$$\boldsymbol{u}(0) = \boldsymbol{u}_0 \tag{4A.1b}$$

and its solution $\boldsymbol{u}(t; \boldsymbol{u}_0) \in \mathbb{R}^N$.

For some time sequence $\{t_n\}$ $(t_1 < t_2 < \cdots < t_n \to \infty)$, a point $\boldsymbol{q} \in \mathbb{R}^N$ satisfying

$$\boldsymbol{q} = \lim_{n \to \infty} \boldsymbol{u}(t_n; \boldsymbol{u}_0) \tag{4A.2}$$

is called an ω-limit point of $\boldsymbol{u}_0$. The whole set of ω-limit points for a given $\boldsymbol{u}_0$ is called the ω-limit set of $\boldsymbol{u}_0$ and denoted by $\omega(\boldsymbol{u}_0)$. ω-limit set is the set of possible states of a solution in $t \to \infty$. Note that the limit is not $t \to \infty$, but $n \to \infty$ for the time sequence $\{t_n\}$. This is because the limit of $n \to \infty$ captures not only the case where $\boldsymbol{u}(t; \boldsymbol{u}_0)$ converges to a single point, but also the case where $\boldsymbol{u}(t; \boldsymbol{u}_0)$ converges to an orbit, which is a set of points. Similarly, for the time sequence $\{t_n\}$ $(t_1 > t_2 > \cdots > t_n \to -\infty)$, the entire set of points satisfying (4A.2) is called the α-limit set of $\boldsymbol{u}_0$ and denoted by $\alpha(\boldsymbol{u}_0)$.

The ω- and α-limit sets are invariant sets of the dynamical system. A positive invariant set S is a set of points in an N-dimensional space such that the solution $\boldsymbol{u}(t)$ remains in the set S as time evolves ($\boldsymbol{u}(t) \in S$, $\forall t > 0$) when the initial value $\boldsymbol{u}(0)$ is taken in the set S ($\boldsymbol{u}(0) \in S$). Similarly, a negative invariant set S' is a set such that $\boldsymbol{u}(t) \in S'$ for all $t < 0$ when the initial value is taken in S' ($\boldsymbol{u}(0) \in S'$). Invariant sets are a generalization of the concept of equilibrium points of dynamical systems to dynamical behaviors with finite dimensional extent in the state space such as periodic orbits. In a rigorous discussion of dynamical systems, it is necessary to define the concept of asymptotic behavior in the limit of time $t \to \infty$. In the dynamics of natural phenomena, however, it is more practical to think of these as referring to behavior after some time has passed.

Mathematical Method 4B (Poincaré-Bendixson Theorem)

It is difficult to rigorously show the existence of periodic oscillatory solutions from local stability analysis. In the case of systems of differential equations in two variables, the Poincaré-Bendixson theorem is one way to show the existence of periodic oscillations. In this section, we introduce an intuitive understanding of how to show the existence of periodic solutions from the Poincaré-Bendixson theorem. For details of the theorem, please refer to Strogatz, "Nonlinear Dynamics and Chaos," etc.

Let a system of ordinary differential equations

$$\begin{aligned} \frac{du}{dt} &= f(u, v) \\ \frac{dv}{dt} &= g(u, v) \end{aligned} \tag{4B.1}$$

on a two-dimensional plan $\mathbb{R}^2$ satisfy the following two conditions.

(i) There exists a closed curve B on the plane $\mathbb{R}^2$, such that the vector fields (du/dt, dv/dt) point toward the interior of B at all points (u, v) on B. That is,

$$\boldsymbol{n}(u,v)\cdot(f(u,v),g(u,v))<0,\ \ \forall(u,v)\in B \tag{4B.2}$$

where $\boldsymbol{n}(u, v)$ is the outward normal vector and dot $(\cdot)$ is the inner product.

(ii) There is only one equilibrium point inside the closed curve B, which is either an unstable focus or an unstable node.

In this case, there exists (a) periodic orbit(s) C inside the closed curve B such that $(u(t), v(t))$ asymptotes at $t \to \infty$ when an initial value is taken inside B. If there are multiple closed curves B, the solution asymptotically approaches one of them depending on the initial value (Fig. 4.B1).

A periodic orbit such that the solution converges at $t \to \infty$ is called a stable limit cycle. It should be noted that this theorem is unfortunately not valid for dynamical systems of three or more dimensions.

In the case of living systems, it is often not too difficult to find a closed curve B such that (i) is satisfied. In biomolecular dynamics, terms that increase the amount of biomolecules, such as the transcription function, are usually upper bounded, while the decomposition term is expressed as an increasing function of the amount of molecules. Therefore, a closed curve consisting of four lines: a line parallel to the v-axis with small u, a line parallel to the v-axis with large u, a line parallel to the u-axis with small v, and a line parallel to the u-axis with large v can be a candidate for B.

Mathematical Method 4C (Differential Equations with Time Delay)

Let us show using an example that periodic oscillation solutions can appear when ordinary differential equations include time delays. Consider a simplified version of (4.7), a differential equation with a time delay in one variable.

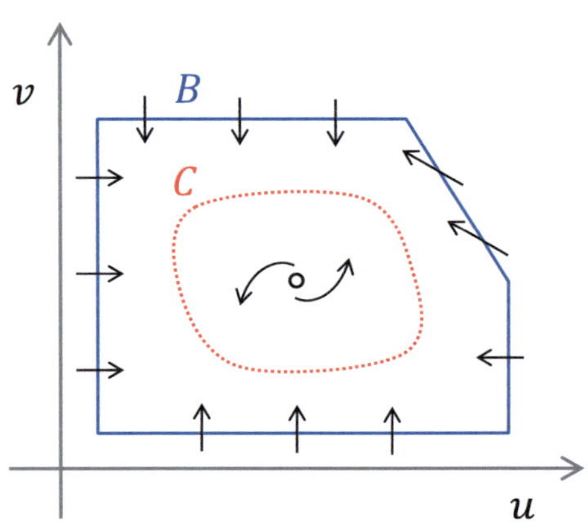

Fig. 4.B1 Schematic diagram of closed curve B and periodic orbit C

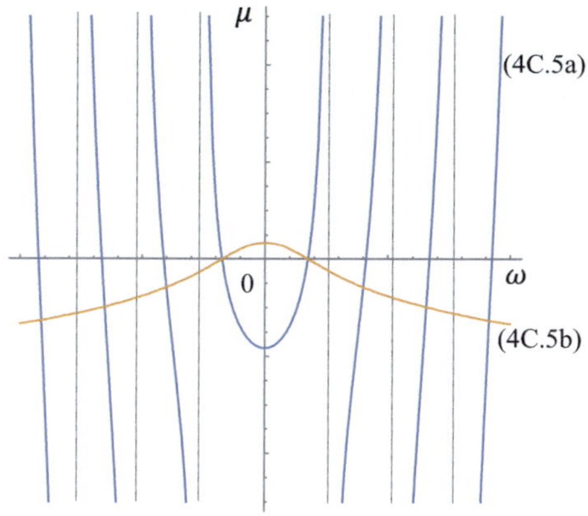

Fig. 4.C1 Two curves determining the real component μ and the imaginary component ω of the eigenvalues. Blue corresponds to (4C.5a) and orange to (4C.5b)

$$\frac{dp(t)}{dt} = \frac{a}{c}f(p(t-\tau)) - bp(t) \tag{4C.1}$$

Assuming that (4.7a) is always at equilibrium, and setting $T_p = 0$, we obtain (4C.1). Let $p(t) = p^*$ be the equilibrium point ($af(p^*)/c - bp^* = 0$) and perform a local stability analysis at this equilibrium point. By setting the deviation from the equilibrium point as $P(t) := p(t) - p^*$, the differential equation with time delay is linearized as follows.

$$\frac{dP(t)}{dt} = \frac{a}{c}f'(p^*)P(t-\tau) - bP(t) \tag{4C.2}$$

Assuming the solution form $P(t) = Ae^{\lambda t}$, we obtain

$$\lambda = \frac{a}{c}f'(p^*)e^{-\lambda\tau} - b \tag{4C.3}$$

Since the eigenvalue λ may be complex, we substitute $\lambda = \mu + i\omega$ into (4C.3) and organize it by real and imaginary components to obtain the following.

$$b + \mu = \frac{a}{c}f'(p^*)e^{-\mu\tau}\cos\omega\tau \tag{4C.4a}$$

$$\omega = -\frac{a}{c}f'(p^*)e^{-\mu\tau}\sin\omega\tau \tag{4C.4b}$$

The real component μ and the imaginary component ω of the eigenvalue $\lambda = \mu + i\omega$ are obtained as the intersection of the following two curves:

$$b + \mu = -\omega\cot\omega\tau \tag{4C.5a}$$

$$(b+\mu)^2 + \omega^2 = \left(\frac{a}{c}f'(p^*)e^{-\mu\tau}\right)^2 \tag{4C.5b}$$

When $\omega = 0$, the right-hand side of (4C.5a) is assumed to be zero.

In Fig. 4.C1, the two curves corresponding to (4C.5a) and (4C.5b) are drawn on the (ω, μ) plane. There are an infinite number of intersections (eigenvalues): the two curves are symmetric with respect to $\omega = 0$, and there always appear pairs of intersections with positive and negative ω of equal absolute value. Among them, the intersection with the maximum μ is the one closest to $\omega = 0$. If $\mu > 0$ at the intersection, this equilibrium point is unstable. Since the complex component is $|\omega| > 0$, the solution is expected to oscillate away from the equilibrium point, resulting in a periodic solution. In the above example, such instability occurs when τ and $af'(p^*)/c$ are large.

Exercise

Exercise 4.1

For the Eq. (4.1), when$a_1 = a_2 = a$ and $d_1 = d_2 = d$, write the differential equations that govern $m(t) := M_1(t) - M_2(t)$ and $r(t) := R_1(t) - R_2(t)$. Also, show that $m(t) \to 0$ and $r(t) \to 0$ as $t \to \infty$. How can you describe the behavior of the per gene (M_1, R_1) and tim gene (M_2, R_2) in this case?

Exercise 4.2

When the assumptions of Exercise 4.1 hold, derive the three-dimensional dynamical system of Eq. (4.2) from Eq. (4.1). Let $\left(\widehat{M}, \widehat{R}, \widehat{P}\right)$ be an equilibrium point of this equation, and find the equations satisfied by $\widehat{M}$, $\widehat{R}$ and $\widehat{P}$.

Exercise 4.3

We study the local stability of the equilibrium point in Exercise 4.2. Show that the characteristic equation satisfied by the eigenvalues is given by:

$$\lambda^3 + A_1\lambda^2 + A_2\lambda + A_3 = 0, \tag{4.9a}$$

$$\begin{aligned} A_1 &= a + c + d + 2b\widehat{R} \\ A_2 &= ac + ad + cd + 2ab\widehat{R} \\ A_3 &= acd + 2sb\widehat{R}\frac{n\left(\widehat{P}/h\right)^{n-1}}{h\left\{1 + \left(\widehat{P}/h\right)^n\right\}^2} \end{aligned}. \tag{4.9b}$$

Exercise 4.4

From the Routh–Hurwitz condition (Mathematical Method 5A), the conditions for the stability of an equilibrium point of a three-variable ordinary differential equation are given by the coefficients A_1, A_2, A_3 of the characteristic equation $\lambda^3 + A_1\lambda^2 + A_2\lambda + A_3 = 0$, i.e. the coefficients A_1, A_2, A_3 must satisfy the inequalities $A_1 > 0$, $A_3 > 0$, and $A_1A_2 - A_3 > 0$. In the characteristic equation obtained in Exercise 4.3, the first and second inequalities always hold, so we determine the conditions under which $A_1A_2 - A_3 > 0$ holds. Show that $A_1A_2 \geq 9acd$ + (positive term) and $A_3 < (2n + 1)acd$ hold. (Hint: Use $x^2 + y^2 \geq 2xy$ to derive the former, and use the condition for equilibrium points (Exercise 4.2) to derive the latter.) From this result, find the necessary conditions that n must satisfy in order for the equilibrium point to be unstable.

Exercise 4.5

For the linearized Eq. (4.4), show that the eigenvalues λ satisfies the characteristic equation:

$$(\lambda + d)^3 + abc = 0. \tag{4.10}$$

Solve this to confirm that the three eigenvalues are:

$$\begin{aligned} \lambda_1 &= -d - \sqrt[3]{abc} \\ \lambda_{2,3} &= -d + \frac{\sqrt[3]{abc}}{2} \pm \frac{\sqrt{3}\sqrt[3]{abc}}{2} i. \end{aligned} \tag{4.11}$$

Using the above results, show that the equilibrium point in (4.4) becomes an unstable focus when $2d < \sqrt[3]{abc}$.

Exercise 4.6

Answer the questions about the following linear ordinary differential equation:

$$\frac{d}{dt}\begin{pmatrix} m \\ p \end{pmatrix} = \begin{pmatrix} -d & -c \\ b & -d \end{pmatrix}\begin{pmatrix} m \\ p \end{pmatrix}, \tag{4.12}$$

where all parameters are positive. Is the equilibrium point (origin) a node or a focus? (3A Mathematical method, see Figs. 3.A1 and 3.A2) Can this equilibrium point produce a Hopf bifurcation?

Exercise 4.7

For Eq. (4.6), find $\partial V/\partial P$ and confirm that it does not depend on M. Similarly, confirm that $\partial V/\partial M$ does not depend on P. Find (M, P) that gives the minimum value of $V(M, P)$.

Exercise 4.8

For Eq. (4.6), show that $dV/dt \leq 0$ using $dV/dt = \partial V/\partial P \cdot dP/dt + \partial V/\partial M \cdot dM/dt$, and $f\left(\widehat{P}\right) - h\left(\widehat{M}\right) = 0$, $g\left(\widehat{M}\right) - k\left(\widehat{P}\right) = 0$. (Hint: Derive $dV/dt = \left\{f(P) - f\left(\widehat{P}\right)\right\}\left\{k(P) - k\left(\widehat{P}\right)\right\} - \left\{g(M) - g\left(\widehat{M}\right)\right\}\left\{h(M) - h\left(\widehat{M}\right)\right\}$.)

References

Bessho Y, Miyoshi G, Sakata R, Kageyama R (2001a) Hes7: a bHLH-type repressor gene regulated by Notch and expressed in the presomitic mesoderm. Genes Cells 6:175–185

Bessho Y, Sakata R, Komatsu S, Shiota K, Yamada S, Kageyama R (2001b) Dynamic expression and essential functions of Hes7 in somite segmentation. Genes Dev 15:2642–2647

Dunlap JC (1999) Molecular bases for circadian clock. Cell 96:271–290

Goldbeter A (2002) Computational approaches to cellular rhythms. Nature 420:238–245

Hardin PE, Hall JC, Rosbash M (1990) Feedback of the Drosophila period gene product on circadian cycling of its messenger RNA levels. Nature 343:536–540

Harima Y, Takashima Y, Ueda Y, Ohtsuka T, Kageyama R (2013) Accelerating the tempo of the segmentation clock by reducing the number of introns in the Hes7 gene. Cell Rep 3:1–7

Hirata H, Bessho Y, Kokubu H, Masamizu Y, Yamada S, Lewis J, Kageyama R (2004) Instability of Hes7 protein is crucial for the somite segmentation clock. Nat Genet 36:750–754

Kurosawa G, Mochizuki A, Iwasa Y (2002) Comparative study of circadian clock models, in search of processes promoting oscillation. J Theor Biol 216:193–208

Lewis J (2003) Autoinhibition with transcriptional delay a simple mechanism for the Zebrafish somitogenesis oscillator. Curr Biol 13:1398–1140

Shimojo H, Isomura A, Ohtsuka T, Kori H, Miyachi H, Kageyama R (2016) Oscillatory control of Delta-like1 in cell interactions regulates dynamic gene expression and tissue morphogenesis. Genes Dev 30:102–116

Takashima Y, Ohtsuka T, González A, Miyachi H, Kageyama R (2011) Intronic delay is essential for oscillatory expression in the segmentation clock. Proc Natl Acad Sci USA 108:3300–3305

Young MW, Kay SA (2001) Time zones: a comparative genetics of circadian clocks. Nat Rev Genet 2:702–715

Chapter 5
Oscillation in Living System II: Cyanobacteria Circadian Rhythms

Abstract In the previous chapter, we discussed several examples of living systems in which periodic oscillations arise from regulatory systems with negative feedback. On the other hand, there are cases in which periodic oscillations are generated from systems that apparently do not have active regulation, such as feedback. Here, we focus on the state transition system of the protein that controls the circadian rhythm of cyanobacteria and try to mathematically clarify the relationship between the structure of biomolecular systems and the dynamics arising from them.

Keywords State Transition · Cyanobacteria circadian rhythm · Conserved quantity · Compartment matrix · Routh · Hurwitz condition

5.1 Cyanobacteria Circadian Rhythms

Cyanobacteria are characterized by their ability to perform photosynthesis and also by their circadian rhythms (Ditty et al. 2003). It is thought that cyanobacteria have a circadian rhythm for adaptive reasons, such as being able to predict changes in the external environment over a 24-hour period and express molecules for photosynthesis in advance. For the molecular mechanism of the cyanobacterial circadian rhythm, three clock genes with critical functions have been identified, which are called *kaiA*, *kaiB*, and *kaiC* (Kondo et al. 1994).

It was initially thought that negative feedback in the regulation of gene transcription might be responsible for creating circadian oscillations even in cyanobacteria. However, a discovery that overturned this assumption was made by the group of Professor Takao Kondo (Nakajima et al. 2005; Tomita et al. 2005). They found that the phosphorylation level of KaiC oscillates in 24 hours simply by mixing purified KaiA, KaiB, and KaiC proteins with ATP in vitro (Fig. 5.1a). In the in vitro experiment, since transcription and translation do not work, the total amount of protein is almost conserved. Furthermore, KaiC phosphorylation oscillations show temperature compensation in vitro, and their periods are almost equal between in vivo and in vitro. Based on the above, the KaiC phosphorylation oscillation is considered to be the essential component of the circadian clock of cyanobacteria.

A. Mochizuki, *Structural Theories for Dynamical Networks in Biology*, Theoretical Biology, https://doi.org/10.1007/978-981-96-6026-1_5

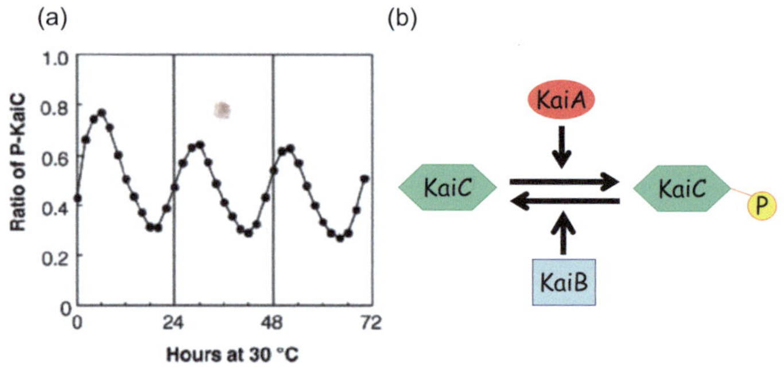

Fig. 5.1 (**a**) Phosphorylation oscillation by the Kai clock protein (from Nakajima et al. 2005). (**b**) Schematic of KaiC phosphorylation regulation; KaiA promotes KaiC phosphorylation, and KaiB suppresses the effect of KaiA

The functions of KaiA, KaiB, and KaiC proteins are summarized in Fig. 5.1b. KaiC, which plays a central role, is a hexameric protein with two phosphorylation sites in each monomer. KaiC has autophosphorylation and dephosphorylation activities, and KaiA promotes the autophosphorylation of KaiC, whereas KaiB suppresses the autophosphorylation of KaiC. Let us directly interpret the schematic diagram in Fig. 5.1b and consider a reversible phosphorylation reaction, in which the forward and backward reactions are promoted by KaiA and KaiB, respectively. In this case, the system would converge to an equilibrium state where the phosphorylated and non-phosphorylated states exist in a certain ratio and would not oscillate. It can be seen that a little more mechanism is needed for periodic oscillations to occur.

The state transition of Kai proteins in cyanobacterial cells during a 24-hour period has been roughly characterized by gel filtration and immunoprecipitation methods (Kitayama et al. 2003; Kageyama et al. 2003, 2006). In the early daytime, KaiC is relatively unphosphorylated, and the majority of KaiC is present as a homohexamer. Eventually, phosphorylated KaiC increases. At this stage, it is thought that KaiA weakly binds to KaiC and phosphorylates many KaiC proteins by repeating binding and disassociation. As phosphorylated KaiC accumulates, the number of KaiA-KaiC complexes gradually increases. Subsequently, accumulation of the KaiA-KaiB-KaiC complex is observed. Thereafter, the KaiC homo-hexamer increases again, and the phosphorylation level decreases. Figure 5.2 summarizes these state transitions, interpreted as a sequence between four states of KaiC: a dephosphorylated state (NP-KaiC), a phosphorylated homo-hexamer (P-KaiC), a small complex (KaiAC), and a large complex (KaiABC).

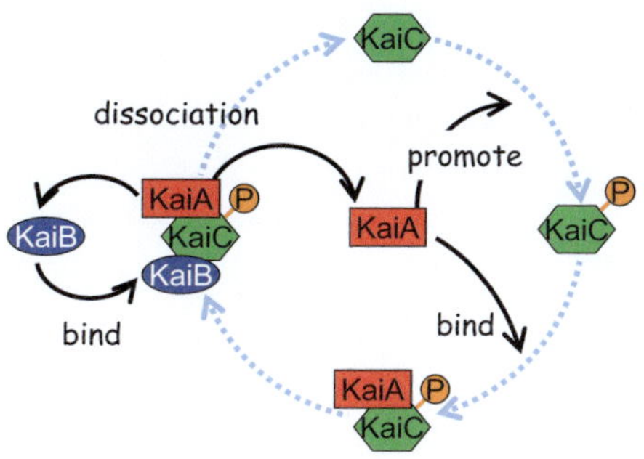

Fig. 5.2 Schematic of the basic model. Based on the observed transition of Kai proteins. The dotted line represents the KaiC state transition reaction, and the solid line represents the regulation of the reaction. The KaiAC and KaiABC complexes are thought to negatively regulate the KaiC phosphorylation reaction by trapping free KaiA

5.2 Mathematical Model of KaiC State Transition System

Imamura et al. were the first to mathematically analyze the effect of "implicit negative feedback" via free KaiA on the KaiC transition system and to clarify the conditions for the occurrence of phosphorylation oscillations (Takigawa-Imamura and Mochizuki 2006). From the state transition network shown in Fig. 5.2, KaiA is involved in two different functions: promoting KaiC phosphorylation by weak binding and complex formation by strong binding. This implies that the system contains negative feedback. That is, as the amount of KaiAC and KaiABC complexes increases, the amount of free KaiA decreases. The decrease in free KaiA is thought to slow down the phosphorylation of KaiC. In other words, there is a negative feedback from the two complexes (KaiAC complex and KaiABC complex) to the phosphorylation process via the amount of free KaiA. An ordinary differential equation model incorporating this effect is given by

$$\begin{aligned}
\frac{du_1}{dt} &= -\frac{k_p g_1 u_1}{k_m + u_1} + k_{dp} u_4 \\
\frac{du_2}{dt} &= \frac{k_p g_1 u_1}{k_m + u_1} - k_{c1} g_2 u_2 \\
\frac{du_3}{dt} &= k_{c1} g_2 u_2 - k_{c2} h u_3 \\
\frac{du_4}{dt} &= k_{c2} h u_3 - k_{dp} u_4
\end{aligned} \tag{5.1a}$$

$$\begin{aligned}
g_i &= a - u_3 - s u_4 - a_i \\
h &= b - u_4
\end{aligned} \tag{5.1b}$$

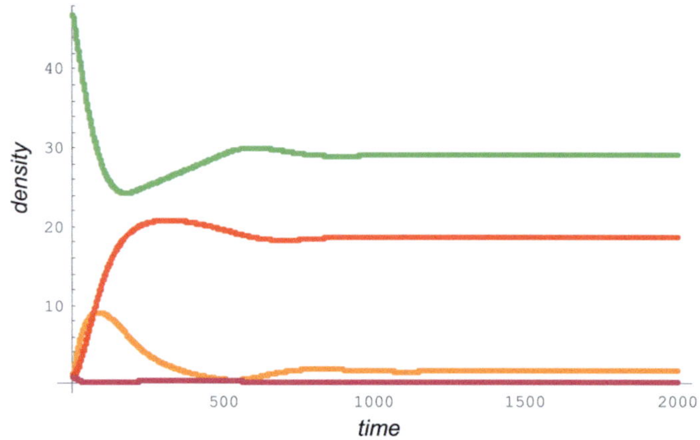

Fig. 5.3 Example of concentration dynamics. Numerically calculated according to Eqs. (5.1a and 5.1b). Green: NP-KaiC, orange: P-KaiC, red: KaiAC, purple: KaiABC concentrations

where u_1, u_2, u_3, and u_4 are the concentrations of NP-KaiC, P-KaiC, KaiAC, and KaiABC, respectively. The phosphorylation reaction of KaiC is represented by the Hill function (Michaelis-Menten equation) with $n = 1$ such that KaiC is the substrate, and KaiA is the regulator. k_p is the rate constant for phosphorylation reaction, and k_m is the Michaelis constant. k_{c1}, k_{c2}, k_{dp} are the rate of phosphorylation, KaiAC formation, and dephosphorylation, respectively. a and b are the total amount of KaiA and KaiB in concentration, respectively. g_1 and g_2 are the effective KaiA concentrations for phosphorylation and KaiAC formation, respectively, and h is the effective KaiB concentration for KaiABC formation. Negative feedback is realized by decrease in g_i and h with increase in complex concentrations (u_3, u_4). a_1 and a_2 indicate the minimum free KaiA concentrations required for phosphorylation and KaiAC formation.

Numerical analysis shows that the system does not oscillate and always converges to a stable equilibrium point. Figure 5.3 shows an example of the time series obtained from numerical simulation of the dynamics. The system temporarily exhibits oscillatory behavior, but the amplitude quickly decays and eventually converges to a steady-state distribution where each state exists in a certain proportion. In this steady-state distribution, each individual molecule cycles through four states, but the concentration of each state in the system is constant with time. Regardless of the choice of the functions or the parameter values, the dynamics of the system always converge to the stationary distribution.

In actual cyanobacteria, the concentration changes with time and oscillates periodically. In other words, it is thought that the steady state becomes unstable and a periodic orbit surrounding it is generated. Therefore, as a necessary condition for oscillations to occur, we study the condition under which the equilibrium point becomes unstable. It is mathematically possible that a stable equilibrium point and a periodic orbit coexist, but this would require the assumption of highly nonlinear function, which is unlikely to be the case in reality.

5.3 Mathematical Analysis

Imamura et al. considered the conditions of the network structure for an equilibrium point of a state-transition system to be unstable using a hypothetical system (Fig. 5.4) (Takigawa-Imamura and Mochizuki 2006). First, they consider a model in which the reaction rate depends only on the concentration of the state before the transition and call it a simple state transition model.

They consider a system in which multiple states of a molecule are possible, and any transition from one state to another is possible. Each variable represents the concentration of each state. The sum of the concentrations of all states does not change over time. The rate of each state transition is assumed to be an increasing function of the concentration of the state before the transition. For example, the reaction rate from state 1 to state 2 is an increasing function of the concentration of state 1, and the reverse reaction is an increasing function of the concentration of state 2. (The reaction rate can be a nonlinear increasing function of the concentration of the state before the transition.) The differential equation for such a reaction system can generally be written as follows

$$
\begin{aligned}
\frac{du_1}{dt} &= \sum_{i\neq 1} r_{i1}(u_i) - \sum_{j\neq 1} r_{1j}(u_1) \\
&\vdots \\
\frac{du_N}{dt} &= \sum_{i\neq N} r_{iN}(u_i) - \sum_{j\neq N} r_{Nj}(u_N)
\end{aligned}
\tag{5.2}
$$

where $r_{ij}(u_i)$ is the reaction rate of the state transition from i to j and is an increasing function of u_i. If the dynamics has an equilibrium point, then the linearized system near the equilibrium point is (Exercise 5.1)

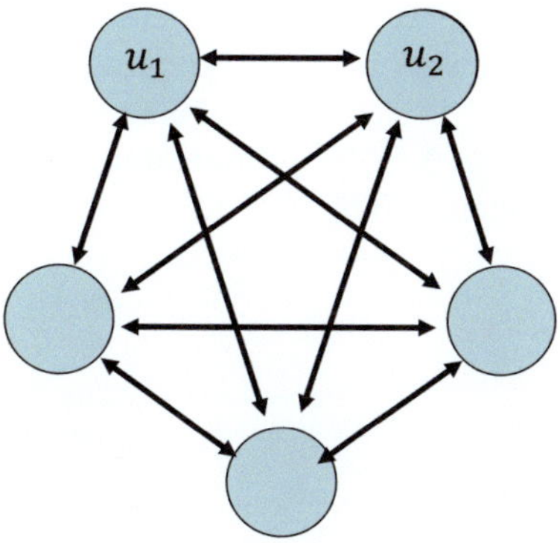

Fig. 5.4 Simple state transition model. The reaction rate is an increasing function of the concentration of the substrate molecules. State transitions are assumed to be possible for all pairs of different molecules

$$\frac{d}{dt}\begin{pmatrix} u_1 \\ u_2 \\ u_3 \\ u_4 \\ u_5 \end{pmatrix} = \begin{pmatrix} -K_1 & k_{12} & k_{13} & k_{14} & k_{15} \\ k_{21} & -K_2 & k_{23} & k_{24} & k_{25} \\ k_{31} & k_{32} & -K_3 & k_{34} & k_{35} \\ k_{41} & k_{42} & k_{43} & -K_4 & k_{45} \\ k_{51} & k_{52} & k_{53} & k_{54} & -K_5 \end{pmatrix}\begin{pmatrix} u_1 \\ u_2 \\ u_3 \\ u_4 \\ u_5 \end{pmatrix} \tag{5.3a}$$

$$\begin{aligned} K_1 &:= k_{21} + k_{31} + k_{41} + k_{51} \\ K_2 &:= k_{12} + k_{32} + k_{42} + k_{52} \\ K_3 &:= k_{13} + k_{23} + k_{43} + k_{53} \\ K_4 &:= k_{14} + k_{24} + k_{34} + k_{54} \\ K_5 &:= k_{15} + k_{25} + k_{35} + k_{45} \end{aligned} \tag{5.3b}$$

where the parameters k_{ij} are all greater than or equal to 0. All diagonal components of the matrix are negative and all non-diagonal components are positive or zero ($k_{ij} \geq 0$). The total amount of the molecules is conserved, so the sum of elements in each column of the matrix is zero.

Here are the important properties defined by the signs of the matrix components. A matrix is called a "compartment matrix," if its non-diagonal components are all non-negative and the sum of elements in each column is non-positive. The largest eigenvalue of a compartment matrix is known to be zero or negative (Mathematical Method 5B). The Jacobian matrix of a simple state transition model is always a compartment matrix. In other words, any equilibrium point of the simple state transition model is always stable and never destabilized.

The simple state transition model assumes that the rate of state transition depends only on the concentration of the state before the transition and is an increasing function of its concentration. In order to destabilize the equilibrium point, we need a mechanism whereby the state transition rate increases or decreases depending on other states. In other words, the system needs feedback. To illustrate this idea more clearly, we consider a new model (feedback model). In this model, we assume a framework of cyclic state transitions in the system. Suppose that the transition from state 1 to state 2 is inhibited by the concentration of another state (state 4 in the case of Fig. 5.5). The differential equation for such a reaction system can generally be written as follows:

$$\begin{aligned} \frac{du_1}{dt} &= r_{N1}(u_N) - r_{12}(u_1, u_p) \\ \frac{du_2}{dt} &= r_{12}(u_1, u_p) - r_{23}(u_2) \\ &\vdots \\ \frac{du_N}{dt} &= r_{N-1,N}(u_{N-1}) - r_{N1}(u_N) \end{aligned} \tag{5.4}$$

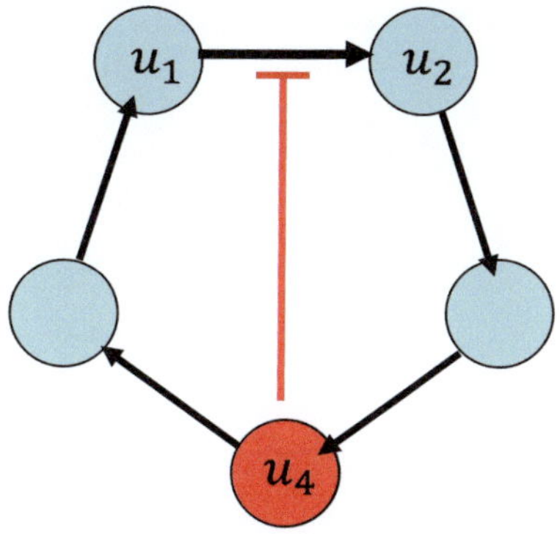

Fig. 5.5 Feedback model. Consider a basic framework with cyclic transitions among N states. (state 1 → state 2 → … → state N → state 1). Assume that the rate of transition from state 1 to state 2 depends on the other states (4 in the figure) and is a decreasing function of its concentration

where $r_{i,\ i\ +\ 1}(u_i)$ is an increasing function of u_i. Only $r_{12}(u_1, u_p)$ depends on two variables and is an increasing function of u_1 and a decreasing function of u_p. In the case of Fig. 5.5, $p = 4$. If an equilibrium point exists, then the linearized systems near the equilibrium point are given by (Exercise 5.1)

$$\frac{d}{dt}\begin{pmatrix} u_1 \\ u_2 \\ u_3 \\ u_4 \\ u_5 \end{pmatrix} = \begin{pmatrix} -k_1 & 0 & 0 & k_f & k_5 \\ k_1 & -k_2 & 0 & -k_f & 0 \\ 0 & k_2 & -k_3 & 0 & 0 \\ 0 & 0 & k_3 & -k_4 & 0 \\ 0 & 0 & 0 & k_4 & -k_5 \end{pmatrix}\begin{pmatrix} u_1 \\ u_2 \\ u_3 \\ u_4 \\ u_5 \end{pmatrix} \tag{5.5}$$

where the parameters $k_i := \partial r_{i,\ i\ +\ 1}/\partial u_i$ and $k_f := -\partial r_{12}/\partial u_p$ are positive. The Jacobian matrix is not a compartment matrix because the (2, 4) component (a non-diagonal component) is negative. We cannot conclude that the equilibrium point is always stable based on the criterion of a compartment matrix.

Therefore, the Routh–Hurwitz condition (Mathematical Method 5A) was used to determine whether the equilibrium point of the feedback model can be unstable or not. The equilibrium point of the system is always stable if the Routh–Hurwitz condition is always satisfied, independent of the parameters (k_i, k_f) in the Jacobi matrix. If the Routh–Hurwitz condition is not satisfied depending on the value of the constants, then the equilibrium point can be destabilized.

The equilibrium point of the system with cyclic state transitions is analyzed for instability by adding a regulation that represses the reaction rate from state 1 to state 2, depending on the concentration of the other states. White circles (○): The system always satisfies the Routh–Hurwitz condition without depending on the parameter value, so any equilibrium of the system is always stable. Black circles (●): Depending on the parameter values, the system does not satisfy the Routh–Hurwitz condition and an equilibrium can be unstable. Due to the symmetry of the cyclic state transition model, the results depend only on the distance between the inhibitor state and the inhibited reaction.

Table 5.1 Stability analysis of the feedback model

System size (N)	Inhibitor u_1	u_2	u_3	u_4	u_5	u_6	u_7	u_8
3	●	○	○	–	–	–	–	–
4	●	○	○	●	–	–	–	–
5	●	○	○	●	●	–	–	–
6	●	○	○	●	●	●	–	–
7	●	○	○	●	●	●	●	–
8	●	○	○	●	●	●	●	●

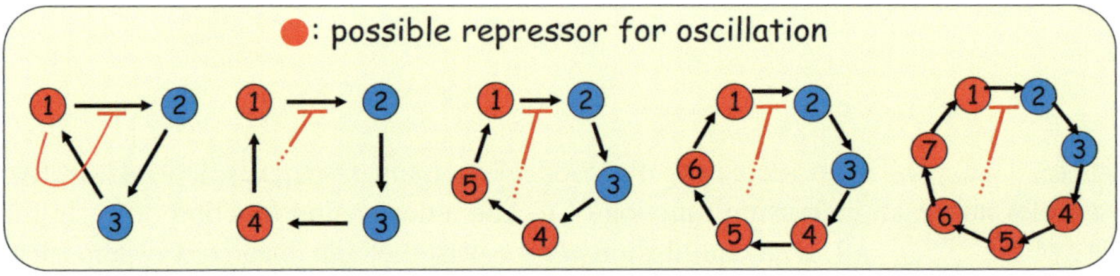

Fig. 5.6 Graphical representation of the results in Table 5.1. In a cyclic state transition model, when the reaction rate from state 1 to state 2 is suppressed, the stability of an equilibrium dependent on the suppressing factor is indicated by the color. If the suppression is from the state indicated in blue, the system will always be stable. If the suppression is from the state indicated in red, the equilibrium point can become unstable

Negative feedback regulation of the state transition model can destabilize the equilibrium point of the system. However, negative feedback alone is not sufficient to destabilize the equilibrium point, and the distance of feedback is very important. The results are summarized in Table 5.1 and visually presented in Fig. 5.6, which shows essentially the same results for systems containing three states up to eight states. The transition from state 1 to state 2 is inhibitory regulated by one of the other states. Linearized system was derived for each system to determine whether the equilibrium point is always stable or can be unstable. The equilibrium point remains stable when suppressive regulation is from state 2 or state 3. For the equilibrium point to be destabilized, suppressive regulation must work from state 4 or later.

From the symmetry of the state transition model, the result should depend only on the distance between the inhibitor state and the inhibited reaction. To summarize the obtained results, let us call the product of an inhibited reaction the first state. For the equilibrium point of the system to be unstable, the suppression from the third or later states must work. Interestingly, this rule of "suppression from the third or later states" is independent of the total system size. In other words, for negative feedback models with system size 3 or larger, we obtain the general rule that for an equilibrium point to be unstable, inhibition from the third or later states is required.

The system can be unstable when the inhibitory regulation acts from state 1 to reaction $1 \rightarrow 2$, but in this case, the equilibrium point does not become the unstable focus, and a periodic oscillation does not appear. In particular, a three-state system can be destabilized only when state 1 is the inhibitor, and no oscillation solution

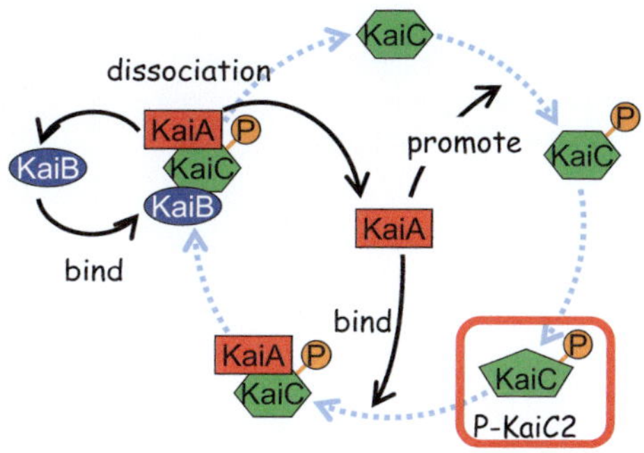

Fig. 5.7 Multiple phosphorylation state model. The existence of a new phosphorylation state (P-KaiC2) is assumed so that there are two states between the reaction subjected to negative feedback (NP-KaiC→P-KaiC) and the inhibiting factors (KaiAC, KaiABC)

appears in this system. A minimum of four states is required for periodic oscillations to occur in the state-transition system with negative feedback. We also analyzed the cases in which inhibitory regulation is applied from multiple states to a single state transition and found that inhibitory regulation from the third state or later is still important.

Let us look at the basic model of the state transition of the Kai protein shown in Fig. 5.2. As explained in the previous section, the first condition of "existence of feedback" is satisfied because negative feedback via the amount of free KaiA from the complexes is working on the phosphorylation reaction. However, the second condition, "the distance between the inhibitor state and the inhibited reaction" is not satisfied. The KaiAC complex, the second state counting from the product of the inhibited reaction is the inhibitor. In this case, the dynamics converge to a stationary distribution that does not oscillate (Fig. 5.7).

In order to satisfy the condition of "suppression from the third or later state," another state must exist after the phosphorylation of KaiC and before the complex formation. Imamura et al. improved the original model (5.1a and 5.1b) and considered a multiple phosphorylation state model, in which there are two phosphorylation states, and the complex is formed after passing through them (Takigawa-Imamura and Mochizuki 2006).

$$\begin{aligned}
\frac{du_1}{dt} &= -\frac{k_p g_1 u_1}{k_m + u_1} + k_{dp} u_4 \\
\frac{du_2}{dt} &= \frac{k_p g_1 u_1}{k_m + u_1} - k_a u_2 \\
\frac{du_a}{dt} &= k_a u_2 - k_{c1} g_2 u_a \\
\frac{du_3}{dt} &= k_{c1} g_2 u_a - k_{c2} h u_3 \\
\frac{du_4}{dt} &= k_{c2} h u_3 - k_{dp} u_4
\end{aligned} \tag{5.6a}$$

$$\begin{aligned}
g_i &= a - u_3 - s u_4 - a_i \\
h &= b - u_4
\end{aligned} \tag{5.6b}$$

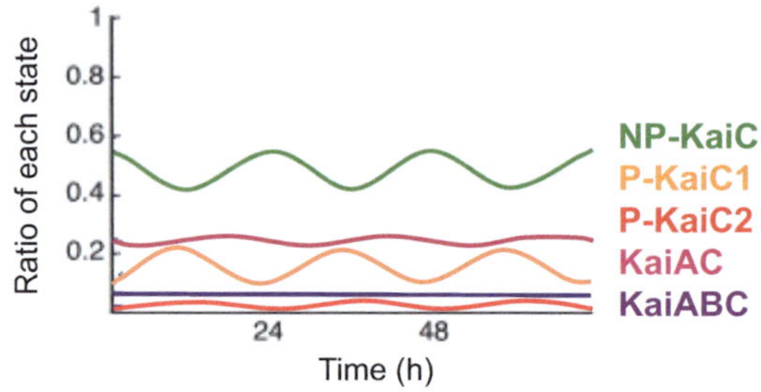

Fig. 5.8 Numerical analysis of the dynamics of the multiple phosphorylation state model

The biological interpretation of this model is as follows: there may be actually multiple phosphorylation states of KaiC, and the complex can be formed only after the transitions through the phosphorylation state 1 (P-KaiC1) and the phosphorylation state 2 (P-KaiC2). Numerical analysis of this model shows that an oscillatory solution does indeed appear, with a clean curve (Fig. 5.8).

The complex formation after phosphorylation and passing through another state can be interpreted as creating a time-delay effect in the transition process from the inhibited reaction to the inhibitor state. Then, one might think that if the transition rate of phosphorylation is sufficiently slow, the same effect can be achieved without increasing the number of states. To confirm whether this intuition is correct, the following analysis was performed.

The multiple phosphorylation state model (5.6a and 5.6b) was analyzed by varying the respective rates of the phosphorylation reaction (k_p) and the phosphorylation interstate transition (k_a) (Fig. 5.9). The black circles (●) indicate that an equilibrium point can be unstable, and the white circles (○) indicate that any equilibrium point is always stable. The results of this analysis indicate that slow transitions between phosphorylation states are important for the instability of the system. In the extreme of the opposite case, i.e., when the transition between phosphorylation states is fast relative to the phosphorylation reaction, the two phosphorylation states can be regarded as a single state. Under such conditions, however, the system converges to a stable equilibrium point.

What is actually the unknown state (P-KaiC2) that exists between phosphorylation and complex formation, as predicted by the mathematical analysis? As explained in the introduction, the KaiC monomer has multiple phosphorylation sites and forms a hexamer, which can have many different phosphorylation states. In a study by the Kondo group, the phosphorylation states of serine 431 and threonine 432 were distinguished, and their changes with time were investigated (Nishiwaki et al. 2007). They clarified that phosphorylation is a multistep process: first, threonine is phosphorylated, then both phosphorylated states accumulate, and gradually only serine phosphorylated state is observed. In KaiC, serine is more easily phosphorylated, but once serine is phosphorylated, threonine is also easily phosphorylated. Such a transition in the phosphorylation state may correspond to the transition among the multiple phosphorylation states predicted by the model.

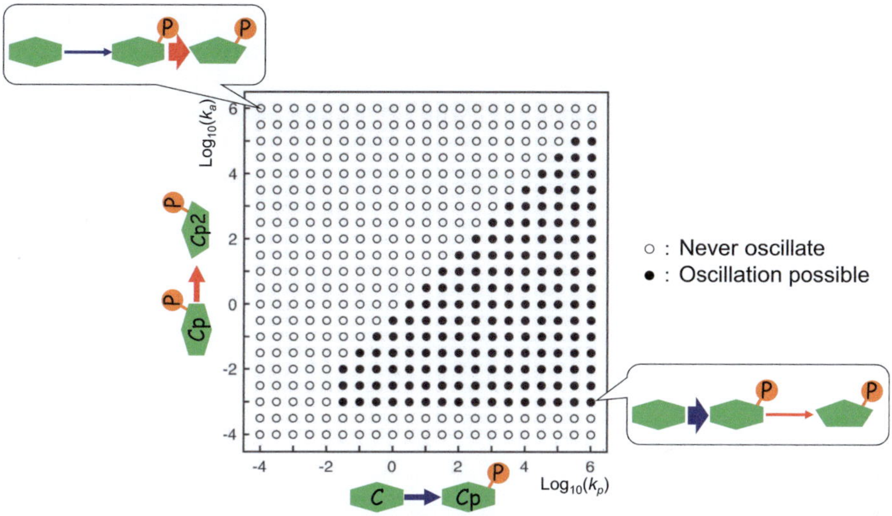

Fig. 5.9 Oscillation conditions on the relative magnitude of the reaction coefficients for the phosphorylation reaction (NP-KaiC→P-KaiC1) and the phosphorylation interstate transition (P-KaiC1→P-KaiC2). Oscillation cannot occur when the phosphorylation reaction (NP-KaiC→P-KaiC1) is slow and the transition between phosphorylation states (P-KaiC1→P-KaiC2) is fast. Oscillations can occur when the transition between phosphorylation states (P-KaiC1→P-KaiC2) is slow and takes a long time

5.4 Summary

In this chapter, we have studied the molecular system that generates periodic oscillations and the physiological function of circadian rhythms. Unlike the examples in the previous chapter, in which negative transcriptional regulation generates periodic oscillations, the example in this chapter is based on a reaction system consisting of a set of state transitions with the conserved quantity of the total amount of molecules. In a system of state transitions, the behavior of the equilibrium point can be studied based on the network structure to some extent. The relationship between the network structure and dynamics will be discussed in a more general way in Chaps. 7, 8 and 9.

Mathematical Method 5A (Routh-Hurwitz Condition)

In Mathematical Method 3A, we discussed the condition that an equilibrium point of a two-dimensional dynamical system is locally stable. A higher-dimensional version of this local stability analysis is known as the Routh-Hurwitz condition. It provides a necessary and sufficient condition for an equilibrium point of a general linear dynamical system to be stable, i.e., for all eigenvalues of the Jacobi matrix to have negative real parts. There are several different mathematically equivalent forms, one of which is presented below. The conditions are given by a set of inequalities consisting of the coefficients of the characteristic equation derived from the Jacobian matrix.

For an n-dimensional dynamical system, the characteristic equation is given by $P(\lambda) = \lambda^n + a_1\lambda^{n-1} + \cdots + a_{n-1}\lambda + a_n = 0$, where a_i, $i = 1, 2, \cdots, n$ are all real numbers. It can also be implicitly assumed that a_n is not identically zero. This is because if $a_n \equiv 0$, then after accepting the obvious solution $\lambda = 0$, we actually should focus on the characteristic equation of the $n - 1$ th order ($a_{n-1} \neq 0$).

Consider the matrix defined below (the Hurwitz matrix)

$$
\begin{gathered}
H_1 = a_1 \\
H_2 = \begin{pmatrix} a_1 & a_3 \\ 1 & a_2 \end{pmatrix} \\
H_3 = \begin{pmatrix} a_1 & a_3 & a_5 \\ 1 & a_2 & a_4 \\ 0 & a_1 & a_3 \end{pmatrix} \\
\vdots \\
H_k = \begin{pmatrix} a_1 & a_3 & a_5 & a_7 & \cdots & a_{2k-1} \\ 1 & a_2 & a_4 & a_6 & \cdots & a_{2k-2} \\ 0 & a_1 & a_3 & a_5 & \cdots & \\ 0 & 1 & a_2 & a_4 & \cdots & \\ \vdots & \vdots & \vdots & \vdots & \cdots & \\ 0 & 0 & 0 & 0 & \cdots & a_k \end{pmatrix}
\end{gathered}
\tag{5A.1}
$$

where a_i is defined only for $1 \leq i \leq n$ and $a_i = 0$ if $i > n$. The necessary and sufficient condition for the real part of all solutions of the characteristic equation $P(\lambda) = 0$ to be negative is given by

$$
\det H_k > 0, k = 1, 2, \cdots, n \tag{5A.2}
$$

Mathematical Method 5B (Perron-Frobenius Theorem and Compartment Matrix)

When components of a matrix satisfy certain conditions, the sign of the real part of the eigenvalues of the matrix can be determined. If the Jacobian matrix of a system satisfies those conditions, the stability of the equilibrium point can be determined from the basic structure of the system alone. Two important theorems are introduced here. The latter is of particular importance in this book.

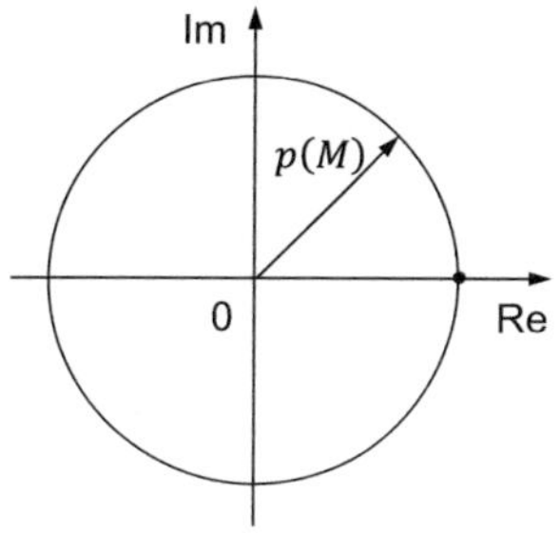

Fig. 5.B1 Equation (5B.4a and 5B.4b) implies that all eigenvalues of M lie within the circle in the figure

1. Matrix Irreducibility

An $n \times n$ real matrix M is called reducible if there exists a permutation $\{i_1, i_2, \cdots, i_n\}$ and a block upper triangular matrix or block lower triangular matrix can be obtained by simultaneous replacement of rows and columns according to the permutation. M is called irreducible if no simultaneous transposition of rows and columns yields a block-triangular matrix.

A matrix M is irreducible if and only if the corresponding directed graph is strongly connected (there is a path from any vertex to any vertex along a directed edge) when M is interpreted as an adjacency matrix (see Chap. 1).

2. Perron-Frobenius Theorem

Suppose that all components of an $n \times n$ real matrix M are greater than or equal to zero and M is irreducible. In this case, there exists a real eigenvalue $p(M)$ with the following properties.

(i) $p(M) > 0$ and all components of the corresponding eigenvector $\boldsymbol{x}$ are positive ($x_i > 0$, $i = 1, \cdots, n$).
(ii) Let $\lambda_i(M)$ be any eigenvalue of M (including complex numbers), then the following holds (Fig. 5.B1).

$$|\lambda_i(M)| \leq p(M), i = 1, \cdots, n \tag{5B.1}$$

In the example of biological phenomena, the transition matrix of discrete-time population dynamics with age structure satisfies the conditions for the Perron-Frobenius theorem. This means that the eigenvalue with the largest absolute value governing the dynamics of the age distribution is always a positive real number, and that the age distribution converges to a stationary distribution rather than oscillating generally after a sufficient time has elapsed. (Note that the meaning of eigenvalues in discrete-time dynamics is different from that in continuous-time dynamics. For details, please refer to a textbook on mathematical ecology.)

3. Compartment Matrix

$n \times n$ real matrix

$$C = \begin{pmatrix} a_{11} & \cdots & a_{1n} \\ \vdots & & \vdots \\ a_{n1} & \cdots & a_{nn} \end{pmatrix} \tag{5B.2}$$

is called a compartment matrix if it satisfies

$$a_{ij} \geq 0, (i \neq j), i, j = 1, \cdots, n \tag{5B.3a}$$

$$\sum_{i=1}^{n} a_{ij} \leq 0, j = 1, \cdots, n \tag{5B.3b}$$

For a compartment matrix C, there exists a real eigenvalue $p(C)$ with the following properties:

(i) $p(C) \leq 0$ and all components of the corresponding eigenvector are greater than or equal to 0.
(ii) For any eigenvalue $\lambda_i(C)$ of C (including the complex case), the following folds (Fig. 5.B2).

$$|\alpha + \lambda_i(C)| \leq \alpha + p(C), i = 1, \cdots, n \tag{5B.4a}$$

where

$$\alpha := \max_i (-a_{ii}) \tag{5B.4b}$$

That is, when the Jacobi matrix is a compartment matrix, the equilibrium point of the system is stable or neutral. The Jacobi matrix for a chemical reaction system (a dynamical system of state transitions) is often a compartment matrix. For example, (5.3a and 5.3b) in the text is a compartment matrix. On the other hand, (5.5) is not a compartment matrix because it has a negative non-diagonal component due to feedback regulation in the system.

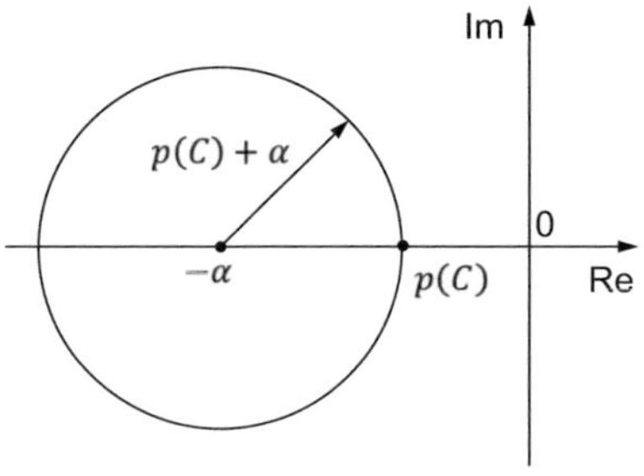

Fig. 5.B2 Equation (5B.4a and 5B.4b) implies that all eigenvalues of C lie within the circle in the figure

Exercise

Exercise 5.1

Suppose that the general differential equation system (5.2) for a simple state transition model has an equilibrium point. Show that the system linearized near the equilibrium point has the form (5.3a and 5.3b). Also, for the general differential equation system (5.4) of the feedback model, assuming that it has an equilibrium point, show that the linearized system near the equilibrium point is of the form (5.5).

Exercise 5.2

Suppose a simple state transition model with (a) molecular inflow and (b) molecular decomposition added, resulting in models where the total amount is not conserved. How can the linearized system around the equilibrium point be written for each case? Also, discuss the stability of each case.

Exercise 5.3

For a four-state feedback model in which the transition from state 1 to state 2 is suppressed by the concentration of state 4, consider the linearized system with simplified parameters

$$\frac{d}{dt}\begin{pmatrix} u_1 \\ u_2 \\ u_3 \\ u_4 \end{pmatrix} = \begin{pmatrix} -k & 0 & 0 & k+h \\ k & -k & 0 & -h \\ 0 & k & -k & 0 \\ 0 & 0 & k & -k \end{pmatrix} \begin{pmatrix} u_1 \\ u_2 \\ u_3 \\ u_4 \end{pmatrix}. \tag{5.7}$$

Determine the stability of the equilibrium point (origin) of this system as follows. Find the characteristic equation satisfied by the eigenvalue λ. Excluding the trivial solution $\lambda = 0$, which is due to the conservation of the total amount, a cubic equation for λ is obtained. Confirm the Hurwitz matrix H_3 is given by

$$H_3 = \begin{pmatrix} 4k & 4k^3+k^2h & 0 \\ 1 & 6k^2 & 0 \\ 0 & 4k & 4k^3+k^2h \end{pmatrix}. \tag{5.8}$$

Also, using the Routh–Hurwitz condition, find the conditions under which the equilibrium point becomes unstable.

Exercise 5.4

For a four-state feedback model in which the transition from state 1 to state 2 is suppressed by the concentration of state 3, consider the linearized system with the simplified parameters in the same way as above

$$\frac{d}{dt}\begin{pmatrix} u_1 \\ u_2 \\ u_3 \\ u_4 \end{pmatrix} = \begin{pmatrix} -k & 0 & h & k \\ k & -k & -h & 0 \\ 0 & k & -k & 0 \\ 0 & 0 & k & -k \end{pmatrix}\begin{pmatrix} u_1 \\ u_2 \\ u_3 \\ u_4 \end{pmatrix}. \tag{5.9}$$

As in Exercise 5.3, use the Routh–Hurwitz condition to discuss the stability of the equilibrium point.

References

Ditty JL, Williams SB, Golden SS (2003) A cyanobacterial circadian timing mechanism. Annu Rev Genet 37:513–543

Kageyama H, Kondo T, Iwasaki H (2003) Circadian formation of clock protein complexes by KaiA, KaiB, KaiC, and SasA in cyanobacteria. J Biol Chem 278:2388–2395

Kageyama H, Nishiwaki T, Nakajima M, Iwasaki H, Oyama T, Kondo T (2006) Cyanobacterial circadian pacemaker: Kai protein complex dynamics in the KaiC phosphorylation cycle in vitro. Mol Cell 23:161–171

Kitayama Y, Iwasaki H, Nishiwaki T, Kondo T (2003) KaiB functions as an attenuator of KaiC phosphorylation in the cyanobacterial circadian clock system. EMBO J 22:2127–2134

Kondo T, Tsinoremas NF, Golden SS, Johnson CH, Kutsuna S, Ishiura M (1994) Circadian clock mutants of cyanobacteria. Science 266:1233–1236

Nakajima M, Imai K, Ito H, Nishiwaki T, Murayama Y, Iwasaki H, Oyama T, Kondo T (2005) Reconstitution of circadian oscillation of cyanobacterial KaiC phosphorylation in vitro. Science 308:414–415

Nishiwaki T, Satomi Y, Kitayama Y, Terauchi K, Kiyohara R, Takao T, Kondo T (2007) A sequential program of dual phosphorylation of KaiC as a basis for circadian rhythm in cyanobacteria. EMBO J 26:4029–4037

Takigawa-Imamura H, Mochizuki A (2006) Predicting regulation of the phosphorylation cycle of KaiC clock protein using mathematical analysis. J Biol Rhythm 21:405–416

Tomita J, Nakajima M, Kondo T, Iwasaki H (2005) No transcription-translation feedback in circadian rhythm of KaiC phosphorylation. Science 307:251–254

Chapter 6
Gene Regulatory Systems and Cell Diversity

Abstract We have shown that various behaviors of organisms emerge from the dynamics of biomolecules. In Chap. 2, we saw that multiple steady states of the gene regulatory system produce different properties of cells. Here, we will further promote this idea. In other words, we will generally consider the diversity of cell types as the diversity of the steady states of a regulatory system. Cell diversity has been considered as a measure of the complexity of multicellular organisms. Therefore, considering the origin of cell diversity also leads us to consider the origin of the complexity of organism's body.

Keywords Cell fate determination · Cell diversity · Multistability · Boolean network · Steady state diversity · Segment genes

6.1 Cellular Properties and Gene Expression Patterns

Many higher organisms have cells with a variety of properties in their bodies, which create structural complexity and biological functions. Complex and functional structures are created over time during development from a single cell, the fertilized egg. Simply put, the function of developmental process can be understood from two major aspects: to produce cells with various properties and to place them in the proper positions. This chapter focuses on the former, the creation of diversity.

Gene regulatory systems producing cell diversity have been experimentally identified in several species. Fig. 6.1 shows the gene regulatory system that is responsible for the specification of the fates of seven types of tissues, including epidermis, brain, nerve cord, muscle, notochord, mesenchyme, and endoderm, during early development of the ascidian from the 8-cell stage to the tailbud stage (Imai et al. 2006). The tailbud embryo is tadpole-like in shape, and the period continues for a while after a very rapid development prior to the period. The figure shows the 81 genes that are active during this period and the regulatory relationships among them. This regulatory network was created through a series of experiments in which each gene was knocked down by morpholino antisense oligonucleotides, and the effects of perturbation were identified by observing changes in gene expression during development.

A. Mochizuki, *Structural Theories for Dynamical Networks in Biology*, Theoretical Biology, https://doi.org/10.1007/978-981-96-6026-1_6

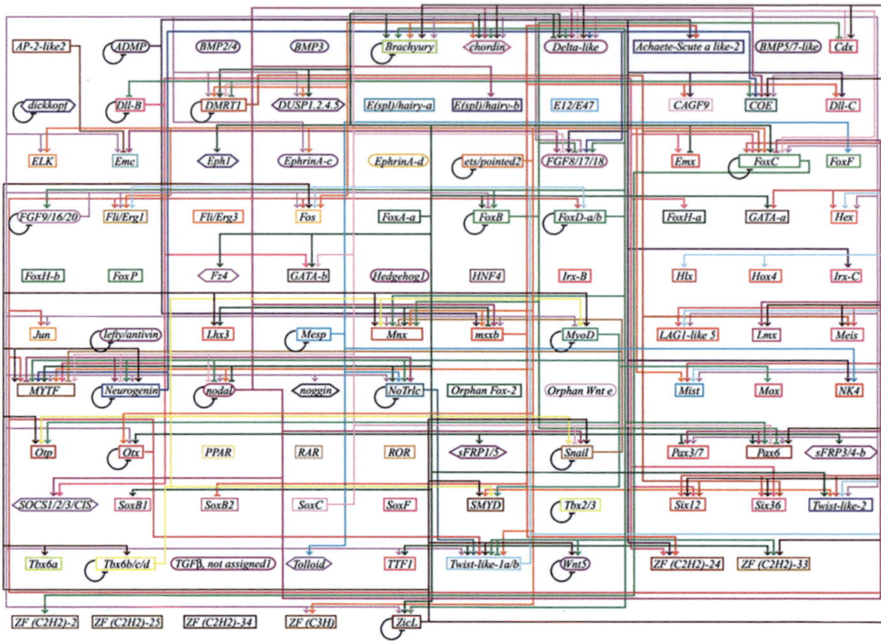

Fig. 6.1 Gene regulatory network of ascidians responsible for the specification of cell fate in early development (Imai et al. 2006). Figure includes 104 vertices. Excluding 23 isolated vertices, it forms a network system with 81 vertices

Fig. 6.2 Conceptual diagram of the increase in diversity of gene expression patterns during development

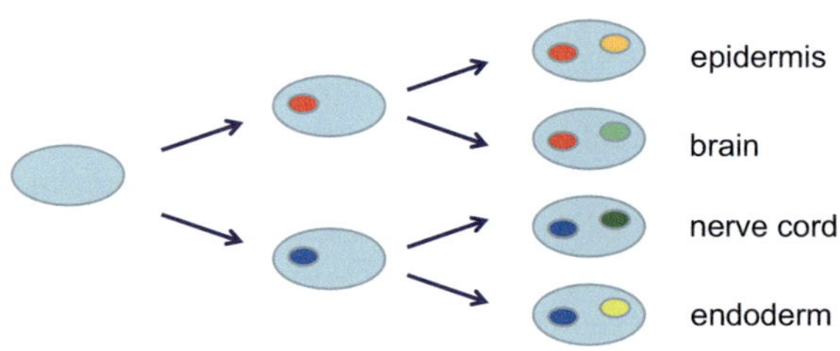

Generally, during development, the number of cells in an embryo increases through cell division, while the diversity of gene expression patterns within the cell population increases (Fig. 6.2). The term "gene expression pattern" is used to indicate which genes are active and which genes are inactive in a cell. Diversification of expression patterns is achieved by the sequential activation of different genes between cells. In the case of ascidians, more than 7 different gene expression patterns are observed at the late gastrula stage, depending on the location of the body. Cells showing different gene expression patterns differentiate into different tissues at later developmental stages. In other words, the diversity of cell types that make up the adult ascidian body is originated in the diversity of gene expression patterns at a specific developmental stage.

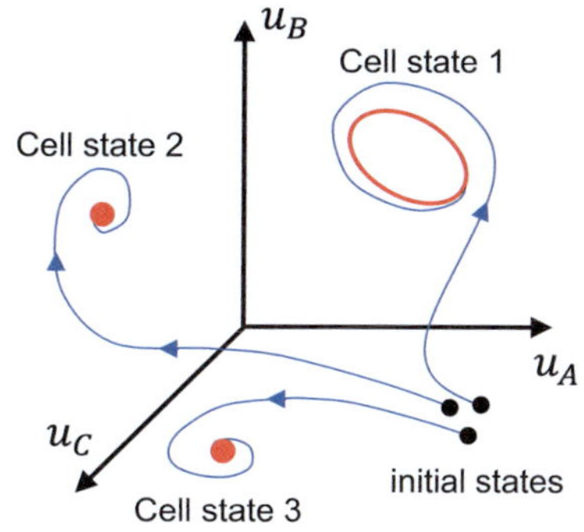

Fig. 6.3 Multistability in dynamics

It should be noted that a variety of gene expression patterns are produced from a single gene regulatory system. Essentially all cells have the same set of genes and the same gene regulatory system. How can we think of a single system producing multiple states?

In general, a nonlinear dynamical system may have multiple goals (ω-limit sets), depending on the initial conditions (Mathematical Method 4A). Figure 6.3 is a conceptual diagram of a dynamical system with two equilibrium points and one periodic oscillation as the ω-limit sets. The existence of multiple ω-limit sets is sometimes referred to as multistability.

It is thought that the multistability of gene expression dynamics may produce the diversity of expression patterns seen in a multicellular organism and, consequently, the diversity of cell types. Multiple different gene expression patterns in the body derived from a single fertilized egg can be understood as a result of biased cytoplasmic factors in a cell before division or cell-cell interactions, leading to the different trajectories of dynamics into different ω-limit sets.

Although this idea is a hypothesis, it is relatively well accepted not only by mathematical biologists but also by experimental developmental biologists. In particular, some cells that arise during development, such as stem cells, maintain their properties even if they are isolated and cultured in vitro. The properties of stem cells may be interpreted as corresponding to the ω-limit set of gene activity dynamics. In Chap. 2, we learned about the systems that generate the two stable equilibrium points. By a similar mechanism, a system that controls cell fate in development may possess a larger number of ω-limit sets.

6.2 Boolean Networks

Here we introduce the simplest mathematical model for the dynamics of gene regulatory systems: the Boolean network. A Boolean network is a dynamical model in which the states are described by discrete binary values and the time evolution proceeds in a discrete manner. Let each of the N genes have one of two

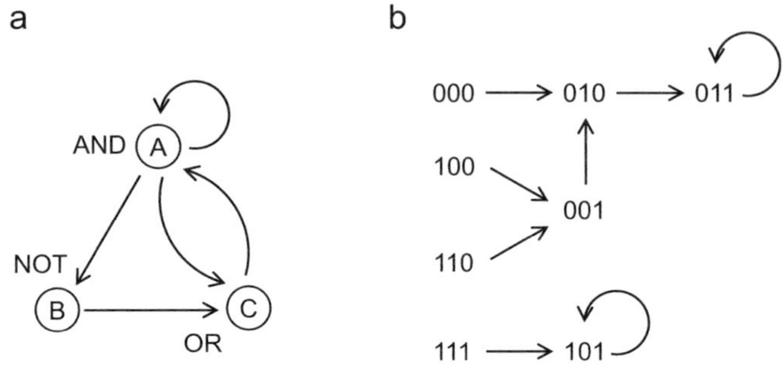

Fig. 6.4 Example of a Boolean network. (**a**) A regulatory network and function of each node. (**b**) State transition graph

states: active (1) or inactive (0). Consider a dynamical system in which N genes update their states simultaneously at discrete points on the time axis.

$$w_n^{t+1} = B_n(\mathbf{w}^t), \quad \left(w_n^t \in \{0, 1\}, n = 1, \cdots, N\right) \tag{6.1}$$

where w_n^t is the state of gene n at time t, $\mathbf{w}^t$ is a state vector in N-dimensional binary space, and B_n is a Boolean function on N-dimensional binary space ($B_n : \{0, 1\}^N \longrightarrow \{0, 1\}$). Since there are two states of each gene, active and inactive, the state of the entire system is represented by a single point in the N-dimensional binary space. In other words, there are 2^N possible states from $(0, 0, \cdots, 0)$ to $(1, 1, \cdots, 1)$. The rule that gives the state of gene n at time $t + 1$ from the states of N genes at time t is B_n. Hereafter, B_n is referred to as the (discrete) regulatory function. Generally, a gene is considered to be expressed under strict conditions, such that the presence of a specific set of transcriptional enhancers and the absence of a specific set of transcriptional repressors allows activation of a gene. The Boolean function is best suited to represent such strict transcription regulation.

Figure 6.4 shows an example of a Boolean network including three genes. Figure 6.4a shows the regulatory relationships among the three genes and the Boolean functions for each gene. The regulatory network shows the dependence between variables but not the specific form of the Boolean functions. In order to uniquely define the dynamics, it is necessary to determine the Boolean function of each node in addition to the regulatory network. Table 6.1 shows the state transitions based on the networks and Boolean functions given in Fig. 6.4a. Each row on the left side of the table shows the state of the system at time t, and each row on the right side of the table shows the state at time $t + 1$.

For example, gene A is regulated by A itself and C, and the regulatory function is assumed to be AND. That is, the state of A at time $t + 1$ is 1 if and only if both A and C are 1 at time t. Gene B is regulated by gene A, and the regulatory function is NOT. Therefore, the state of gene B at time $t + 1$ is the inversion of the state of gene A at time t. Gene C is regulated by genes A and B, and its regulatory function is

Table 6.1 State transition table corresponding to the Boolean network in Fig. 6.4a

t			$t + 1$		
A	B	C	A	B	C
0	0	0	0	1	0
0	0	1	0	1	0
0	1	0	0	1	1
0	1	1	0	1	1
1	0	0	0	0	1
1	0	1	1	0	1
1	1	0	0	0	1
1	1	1	1	0	1

OR. Therefore, the state of C at time $t + 1$ is 1 in the row where either A or B is 1 at time t.

By selecting one state at time t from the left $8(=2^3)$ rows of the table and looking at the right side of the same row, the state at time $t + 1$ after the state transition is obtained. By repeating this process, a series of state transitions, i.e., time evolution, can be determined. The state of the system is represented by a point in a three-dimensional binary space, i.e., a three-digit binary number. Figure 6.4b shows a state transition diagram based on this Boolean network, in which the possible states are limited as the time evolution progresses. In this figure, 011 and 101 have arrows returning to themselves (self-loops). These two states are steady states that are maintained once they are reached in the dynamics. In the state transition table, the rows with equal N digits numbers on both sides of the table indicate the steady states.

Unlike a dynamical system in a continuous state, in a Boolean network, there is no "unstable equilibrium point" for which the solution moves away when it deviates slightly. In other words, all equilibrium points and periodic solutions are "stable." Perhaps for this reason, in the context of Boolean networks, the expression "steady state" is often used rather than equilibrium point.

6.3 Drosophila Segmentation Genes

One of the biological phenomena that have been greatly elucidated by molecular genetics is the regulation of gene expression and cell fate determination in Drosophila. In Drosophila, anterior-posterior axial fate decisions are based on a stepwise process in which genes expressed earlier in a wider range induce the expression of new genes in a narrower range later in development. First, at the time of fertilization, some genes called "maternal effect genes" create a concentration gradient along the anterior-posterior axis. Based on this information, each of the "gap genes" is activated in only a part of the embryo. In the early stages of Drosophila development, nuclear division proceeds without cytoplasmic division, and the nuclei migrate near the surface of the embryo to form a syncytial blastoderm. Specific gap genes are expressed depending on the protein concentration of maternal effect genes at the

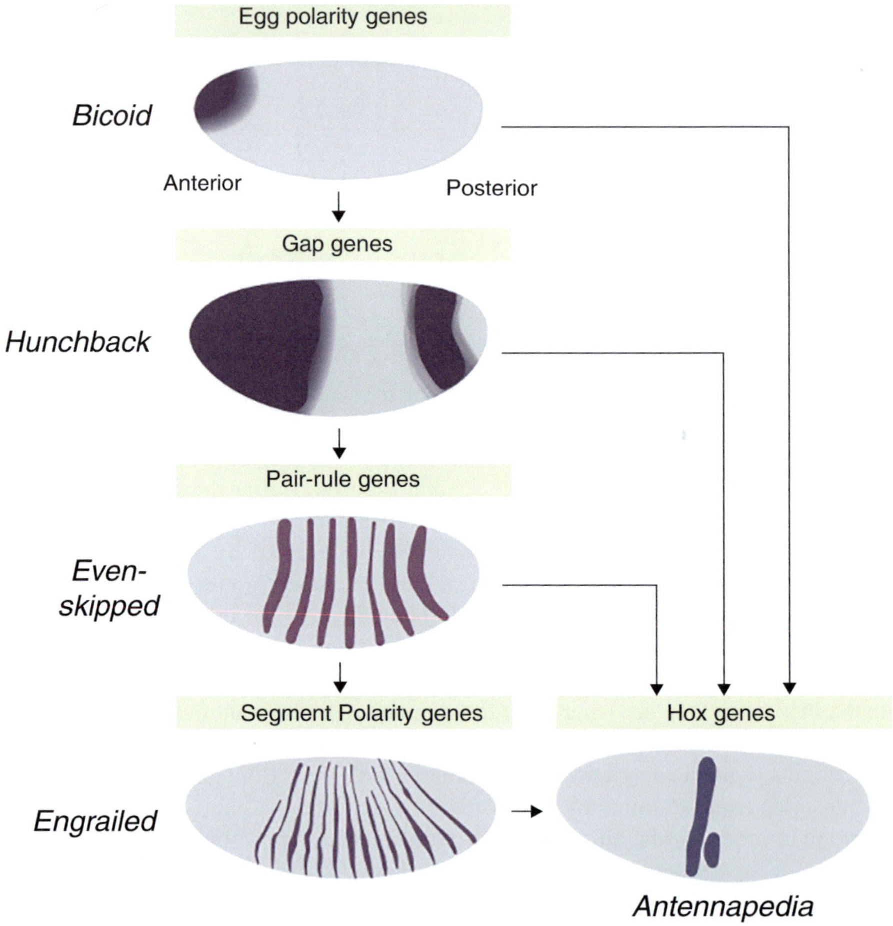

Fig. 6.5 Stepwise regulation of the expression of the Drosophila segmentation gene. (Adapted from Albert et al. 2017, Chapter 21, Figs. 21 and 20)

location of each nucleus. Based on the expression distribution of the gap genes, the expression of the "pair rule genes" is determined, and based on the expression distribution of the pair rule genes, the expression of the "segment polarity genes" is determined. In this way, the expression distribution is progressively subdivided, and differences in cell properties along the anterior-posterior axis of the embryo are created. During the period of segment polarity gene expression, the cell membrane is formed from the surface to the interior, enveloping the nucleus and forming the cellular blastoderm (Fig. 6.5).

Albert and Othmer developed a Boolean network model to study the dynamics of segmentation genes during early development in Drosophila, particularly focusing on the regulation of segment polarity genes based on the expression of pair-rule genes (Albert and Othmer 2003). During Drosophila development, a repeating

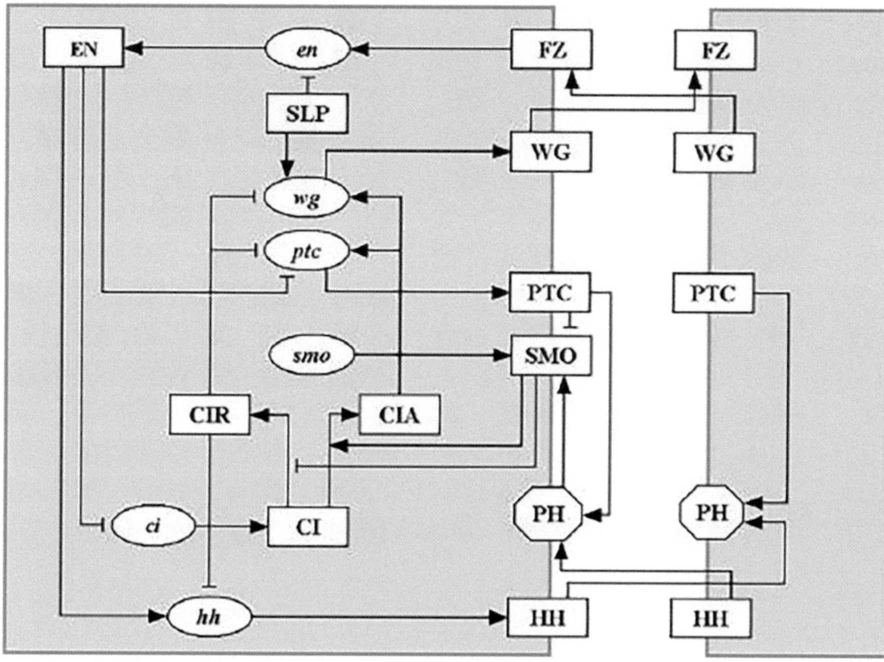

Fig. 6.6 Segmental polarity genes involved in anterior-posterior axis formation in Drosophila and the regulation within and between them. Regulation between adjacent cells is also included. Ovals represent mRNAs, squares represent proteins, and octagons represent protein complexes. SLP is a constant that depends only on cell location. FZ also does not change in time and activates the expression of en when WG is activated in the neighboring cells. (Albert and Othmer 2003)

structure called a parasegment appears along the anterior-posterior axis, and three of these parasegments were subjected to the study. Assuming that each parasegment contains four cells, they considered a structure of 12 cells arranged in one dimension. They focused on six segment polarity genes and considered a network system containing a total of 15 nodes, each consisting of mRNA concentration, protein concentration, and complex concentration among several proteins. One of these genes, the pair-rule gene SLP, is assumed to be continuously expressed in a specific cell within each parasegment. The 12 cells each have 14 variables, making the Boolean network with a total of 168 variables. Some of the regulation between vertices is considered to work within cells, and some to work across cells (Fig. 6.6).

The expression pattern observed in wild-type embryo (Fig. 6.7a) was given as the initial state, and the dynamics of the Boolean network were calculated to obtain the steady-state expression pattern (Fig. 6.7b). The expression pattern obtained by the model was consistent with the actual expression pattern of wild-type development. Note that the expression of all 12 cells is included in one steady state. Similarly, excluding specific genes from the network model reproduced the expression patterns of the mutants. The analysis of the model also allows for an exhaustive study of the steady states. It was found that there are only six possible steady states.

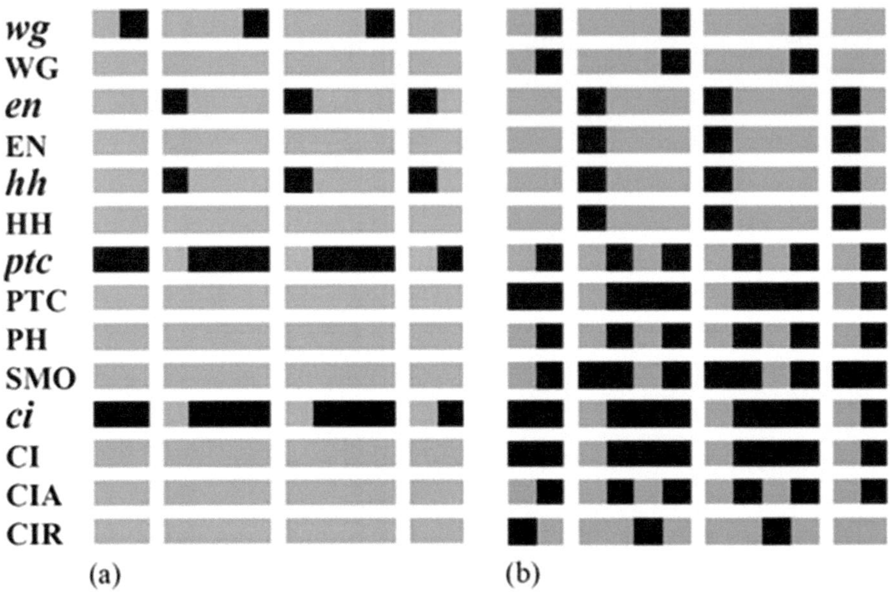

Fig. 6.7 Expression patterns of segment polarity genes calculated by the model. The left side of the figure indicates the anterior side, the right side the posterior side, and the gray area corresponds to the parasegment. (**a**) Stage 8 expression pattern given as the initial state. (**b**) Expression pattern obtained as the steady state. The expression patterns are consistent with those observed in stages 9–11 of actual wild-type embryos. (Albert and Othmer 2003)

This is very small compared to the number of variables in the system (168) and the number of all states (2^{168}). Half of them are patterns that have been actually observed in wild-type development or mutants.

6.4 Exhaustive Analysis of Boolean Networks

Intuitively, it seems that the larger the number of genes in a system, the more complex produced dynamical behaviors are. It has been argued that a large number of genes (N) is the origin of the complexity of the behaviors of Boolean networks. Kauffman, for example, predicted that the number of attractors (ω-limit sets) and cell states would increase as the system size N increased (Kauffman 1969, 1993). However, some studies have come to question the results (Samuelsson and Troein 2003; Drossel et al. 2005).

By considering the state transition table as illustrated in Table 6.1, a discussion covering all Boolean networks is possible. By determining the right-hand side of the state transition table, i.e., setting each of the $N \times 2^N$ components to 0 or 1, the dynamics of an N-vertex Boolean network is uniquely determined. In other words, in the case of N variables, there are $2^{N \times 2^N}$ possible Boolean dynamical systems, including the variety of Boolean functions. It is also possible to discuss how often Boolean networks with a specific property may appear by probabilistic arguments based on this setting.

In the following, we calculate the expected number of steady states of randomly generated Boolean networks (Mochizuki 2005). The cellular states are considered to correspond to ω-limit sets that include also periodic solutions, but for the sake of simplicity, we will first focus on the steady states only. Intuitively, the analysis here corresponds to constructing a Boolean network randomly, recording the number of steady states of the system, and determining the average of the recorded number of steady states after repeating this process many times. Let the probability of assigning 0 or 1 to each component of the state transition table be $1 - p$ and p, respectively. This is to show the generality of the conclusion, and it is also acceptable to set the probability p to 1/2.

In the state transition table, a row in which the N-digit binary numbers are equal on both sides indicates a steady state. Therefore, the probability for each state to be a steady state can be calculated. For example, the state at the first row of the table $(0, 0, \cdots, 0)$ (all genes are inactive) is a steady state with probability $(1 - p)^N$.

Each state becomes a steady state independently from the other. When each state becomes a steady state, its contribution to the total number of steady states is naturally 1. Therefore, the expected value of the number of steady states is given by

$$E_{ss} = 1 \times P_s(00\cdots0) + 1 \times P_s(00\cdots1) + \cdots + 1 \times P_s(11\cdots1) \tag{6.2}$$

where $P_s(S)$ is the probability that state S is a steady state.

Recall that there are $\binom{N}{k}$ N-digit binary numbers that contain k elements with value 1. Where $\binom{N}{k}$ is the binomial coefficient. From the binomial theorem, we obtain:

$$\begin{aligned} E_{ss} &= \binom{N}{0}(1-p)^N + \binom{N}{1}p(1-p)^{N-1} + \cdots + \binom{N}{N}p^N \\ &= \sum_{k=0}^{N}\binom{N}{k}p^k(1-p)^{N-k} = 1 \end{aligned} \tag{6.3}$$

The expected number of the steady states of a randomly constructed Boolean network is 1. Note that in addition to the small and special value of 1, it does not depend on the major parameters. Specifically, it does not depend not only on the number of genes N, but also on the activation rate p nor the number of interactions between genes.

Figure 6.8 intuitively explains why the number of steady states does not depend on the number of genes N. If we consider the expression of a gene as binary, there are a total of 2^N possible expression states. As the number of genes increases, this

Fig. 6.8 Intuitive explanation of why the number of steady states in a randomly constructed system is independent of the number of genes

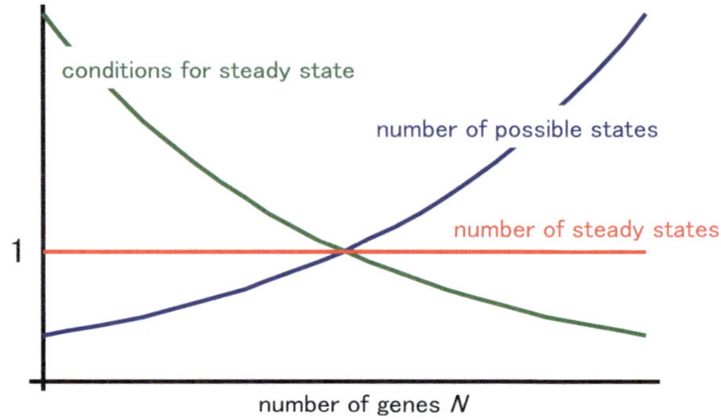

number will of course increase exponentially. However, the condition for each state to become a steady state grows more stringent with the increase in the number of genes N. This is because all N genes must satisfy the condition for a steady state, i.e., the values on the left and right sides of the state transition table must match. The effect of the increase in the number of states exactly cancels out the effect of the conditions to be more stringent, and as a result, the effect of the number of genes disappears. Therefore, the average number of steady states does not change even if the number of genes increases.

A similar approach can be used to obtain a distribution of the number of steady states (Mochizuki 2005). The analysis corresponds to randomly constructing a Boolean network, recording the number of its steady states, and obtaining a frequency distribution of the numbers. For simplicity, let us assume that the activation rate of a gene, i.e., the probability p of assigning 1 to each component of the state transition table, is 1/2. Each of the 2^N possible states has an equal probability of being a steady state, $1/2^N$, as discussed above. Thus, the number of steady states in a random Boolean network is given by the binomial distribution

$$P_{ss}(X=z)=B\left(2^N,\frac{1}{2^N}\right)=\binom{2^N}{z}\left(\frac{1}{2^N}\right)^z\left(1-\frac{1}{2^N}\right)^{2^N-z} \tag{6.4}$$

Figure 6.9 shows the distribution of the number of steady states. The change in the shape of the distribution saturates when the number of genes N is about 5, and thereafter the shape of the distribution is almost independent of N. All distributions have a rapidly decreasing function for $X > 1$. For any value of N, most gene regulatory systems have only a few steady states (0 to 2), and those that produce many steady states are very rare. If a gene regulatory system in a real organism has a large number of steady states, it is located at a place with a large value on the horizontal axis in Fig. 6.9. It is a very rare regulatory system, extremely biased from the average.

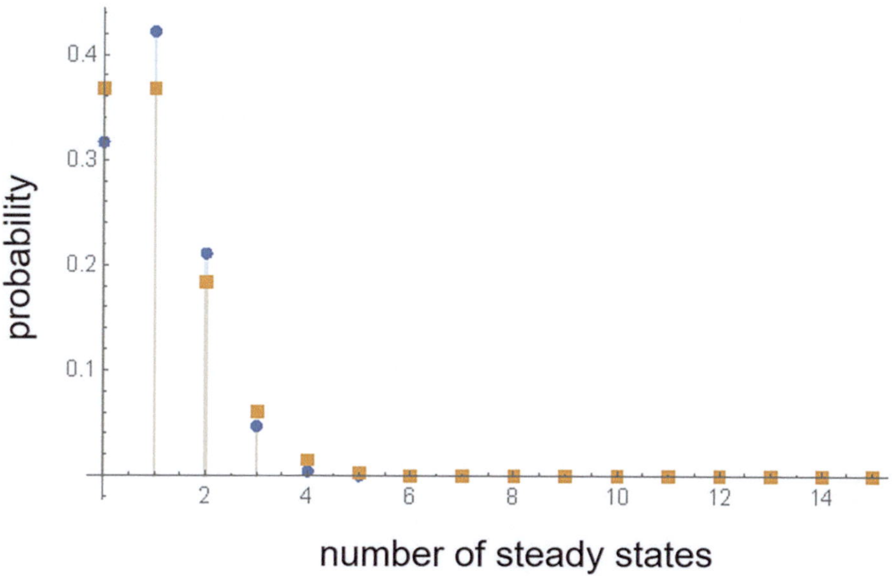

Fig. 6.9 Probability distribution of the number of steady states in a randomly constructed network. Blue: number of genes $N = 2$, orange: number of genes $N = 10$. The shape of the distribution changes little as the number of genes N increases. The expected value of the number of steady states is 1 for all distributions

6.5 Network Structure and Cell States

In a Boolean network, the number of genes is not a direct factor to increase the number of steady states. On the other hand, in real so-called higher organisms, a large number of cell types is considered necessary to produce complex body structures. How should we understand this gap? One may claim that many of the cell types in an organism's body may be realized by periodic oscillations rather than by steady states of gene expression dynamics. And if we consider the ω-limit set of the Boolean network, including periodic solutions, the number of such sets may grow in dependence on the number of genes. However, as the expression "gene expression pattern" suggests, it seems that the sustained expression of genes, i.e., the diversity of steady states, creates the diversity of cell states in real organisms.

If organisms with complex structures have evolved to increase the number of steady states of gene expression, this could not be achieved simply by "increasing the number of genes." The number of genes in various organisms has been studied and estimated. The number of genes in humans is estimated to be about 20,000 or more as a result of the Human Genome Project (International Human Genome Sequencing Consortium 2004). This number is almost the same as that of Drosophila or C. elegans (Harrison et al. 2002). It has been pointed out with large surprise that organisms with complex structures do not necessarily have many genes. However, this fact may not be surprising if one understands the results of this chapter. One cannot discuss the diversity of gene expression patterns only based on the number of genes.

What, then, creates the diversity of gene expression patterns in the actual gene regulatory systems of living organisms? Figure 6.9 shows that the probability of a system with a large number of states is very low in randomly created Boolean networks. In other words, the regulatory systems in real organisms may be extremely biased and rare, unlike those chosen at random.

Recently, we obtained more advanced results on the expected number of steady states of Boolean networks, using the method introduced in Sect. 6.5 of this chapter. It shows that even under a fixed choice of network structure, the expected number of the steady states still remains 1, and the network structure is not a sufficient condition for increasing the number. The intuitive explanation is that for the average steady states to be greater than 1, the regulation on the paths that make up the feedback must be positively biased. In other words, we found that in addition to the network, the regulatory function must also be chosen (Mori and Mochizuki 2017).

At the beginning of this chapter, we considered the two functions of biological developmental process: creating cellular diversity and placing them in the proper positions. The discussion in this chapter can be said to focus only on the former (creation of diversity), leaving the latter (proper spatial arrangement) for later. In actual developmental processes, cell fates may be determined by cell-cell interactions including secretory signaling and mechanical effects between contact cells, and these spatial interactions may contribute to the diversity of cell types.

6.6 Summary

In this chapter, we attempted to understand the diversity of cell types in multicellular organisms as the dynamical behavior of the gene regulation system. A discrete-time discrete-state dynamics, the Boolean network, was used to improve the outlook. Logical conclusions were obtained by exhausting all possible systems using probability theory. The result was that the number of genes is not a direct causal factor of the diversity of the steady states. One of the roles of theory is to provide logical verification for a popular but intuitive belief.

Mathematical Method 6A (Discrete and Continuous Dynamics)

It is relatively easy to obtain general insights into the gene regulatory system by using Boolean networks. However, one should be careful in drawing conclusions using Boolean networks, which are discrete-time discrete-state models. In the

Table 6.A1 State transition table for mutually repressive genes A and B

t		$t+1$	
A	B	A	B
0	0	1	1
0	1	0	1
1	0	1	0
1	1	0	0

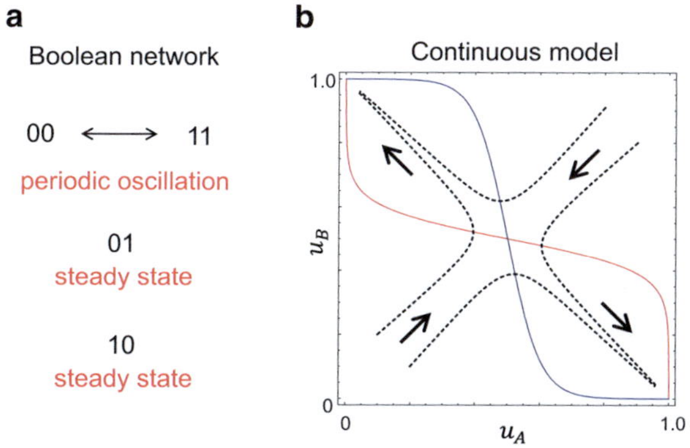

Fig. 6.A1 Dynamics of a Boolean network and a continuous model corresponding to the state transition on Table 6.A1. (**a**) In the case of the Boolean network, in addition to the two steady states (0, 1) and (1, 0), there exists an oscillatory solution with period 2 between state (0, 0) and state (1, 1). (**b**) Continuous model. The differential equations $\dot{u}_A = 1/(1 + \text{Exp}[-20(u_B - 0.5)]) - u_A$, and $\dot{u}_B = 1/(1 + \text{Exp}[-20(u_A - 0.5)]) - u_B$ are used. The red and blue curves are nullclines for $\dot{u}_A = 0$ and $\dot{u}_B = 0$, respectively

ordinary differential equation model, gene expression has a continuous value and changes continuously with time, as described in the previous chapters. The dynamics of the two models may be different, even if the same regulatory system and similar functions are used. This is especially the case for periodic behaviors. Examples are shown in Table 6.A1 and Fig. 6.A1.

Figure 6.A1 shows the dynamics in a Boolean network and an ordinary differential equation model, respectively, using the state transition table in Table 6.A1. In the case of the Boolean network, in addition to the two steady states 01 and 10, the oscillation with period 2 between 00 and 11 is the stable solution. In the ordinary differential equation model, on the other hand, there are stable equilibrium points $(0, 1)$ and $(1, 0)$ corresponding to 01 and 10, but no periodic solution. An unstable equilibrium point (saddle point) exists at the center $(0.5, 0.5)$ of the state space shown in the figure.

In the study of Boolean networks, attractors including both steady states and periodic oscillations are sometimes the subject of analysis. However, it is possible that behaviors identified as periodic oscillations in a Boolean network do not appear in the corresponding continuous-time continuous-state model.

Exercise

Exercise 6.1

For the regulatory network shown in Fig. 6.4a, consider a Boolean network with the regulation function OR for vertex A and the regulation function AND for vertex C. Construct a state transition table and draw a state transition graph.

Exercise 6.2

Consider the Boolean network for the regulatory network shown below. Since each vertex receives only one regulation edge, it is not necessary to assume a regulation function. Construct a state transition table and draw a state transition graph. In the case of a continuous dynamical system, we showed in Chap. 4 that the corresponding regulation system does not exhibit periodic oscillation. Compare this result with the case of capturing dynamics using a Boolean network.

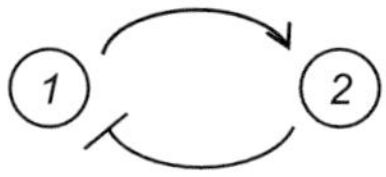

Exercise 6.3

For the case where the number of genes is $N = 3$, how many Boolean network systems are possible, including the possibility of Boolean functions?

Exercise 6.4

For a Boolean network with $N = 3$ genes, assign 0 or 1 to each component on the right-hand side of the state transition table with probabilities of $1 - p$ and p, respectively. Calculate the probability that each state will become a steady state. Also, calculate the expected number of the steady states based on this.

Exercise 6.5

Consider a Boolean network in which the expression of gene n at time $t + 1$, w_n^{t+1}, depends on the expression at time t, w_n^t. Construct the state transition table probabilistically as follows. For each component on the right side of the state transition table, when $w_n^t = 0$, give $w_n^{t+1} = 0$ with probability q and $w_n^{t+1} = 1$ with probability $1 - q$, and when $w_n^t = 1$, give $w_n^{t+1} = 1$ with probability q and $w_n^{t+1} = 0$ with probability $1 - q$. Assuming that the number of genes is $N = 3$, calculate the probability that each state becomes a stationary state. Also, calculate the expected number of steady states.

References

Albert R, Othmer HG (2003) The topology of the regulatory interactions predicts the expression pattern of the segment polarity genes in Drosophila melanogaster. J Theor Biol 223:1–18

Bonner JT (1988) The evolution of complexity by means of natural selection. Princeton University Press, p 276

Davidson EH et al (2002) A genomic regulatory network for development. Science 295:1669–1678

Drossel B, Mihaljev T, Greil F (2005) Number and length of attractors in a critical Kauffman model with connectivity one. Phys Rev Lett 94:088701

Harrison PM, Kumar A, Lang N, Snyder M, Gerstein M (2002) A question of size: the eukaryotic proteome and the problems in defining it. Nucleic Acids Res 30:1083–1090

Harvey I, Bossomaier T (1997) Time out of joint: attractors in asynchronous random boolean networks. In: Husbands P, Harvey I (eds) ECAL 1997. Proceedings of the fourth European conference on artificial life. MIT Press, pp 67–75

Imai KS, Levine M, Satoh N, Satou Y (2006) Regulatory blueprint for a chordate embryo. Science 312:1183–1187

International Human Genome Sequencing Consortium (2004) Finishing the euchromatic sequence of the human genome. Nature 431:931–945

Kauffman SA (1969) Metabolic stability and epigenesis in randomly constructed genetic nets. J Theor Biol 22:437–467

Kauffman SA (1993) The origines of order: self-organization and selection in evolution. Oxford University Press, p 732

Mochizuki A (2005) An analytical study of the number of steady states in gene regulatory networks. J Theor Biol 236:291–310

Mori F, Mochizuki A (2017) Expected number of fixed points in Boolean networks with arbitrary topology. Phys Rev Lett 119:028301

Samuelsson B, Troein C (2003) Superpolynomial growth in the number of attractors in Kauffman networks. Phys Rev Lett 90:098701

Chapter 7
Structure and Dynamics of Regulatory Networks

Abstract With the development of life sciences, information on interactions between biomolecules has accumulated, and these interactions are showing such a complex aspect that they are called networks. It is thought that the dynamics that arise from complex network systems are the essence of biological activity. For networks that have been experimentally identified, clarifying the dynamics that arise from them is one of the most important issues in current biology. On the other hand, networks do not have sufficient information to determine their dynamics. They do not provide information of mathematical formulas or parameter values, that are necessary to determine the details of dynamics of systems. However, even if the dynamics themselves are not determined, there are important aspects of dynamical properties that can be extracted from the network. In this chapter, we will introduce a new mathematical theory that uses only information of regulatory networks to understand the behavior of biological systems. Namely, key factors for observing/controlling the whole dynamical system are determined from network structure alone. We also show an application of the theory to a real biological system, a gene regulatory network for cell-fate specification in ascidian. We demonstrate that the system was completely controllable by experimental manipulations of the key factors identified by the theory from the information of network alone.

Keywords Regulatory network · Structural theory · Linkage logic · Feedback vertex set · Determining nodes · Human circadian rhythm · Ascidian cell specification · Controllability · Observability

7.1 Regulatory Networks in Organisms

In the history of biology, it has become clear that a large number of biomolecules are involved in various biological phenomena and that they have complex regulatory relationships with each other. For example, in the previous chapter, we introduced the gene regulatory network responsible for the specification of seven types of tissues, or cell fate specification, in the early development of the ascidian (Fig. 7.1) (Imai et al. 2006). Let's review what we understood in the previous chapter. In general, during the process of development, the diversity of gene

A. Mochizuki, *Structural Theories for Dynamical Networks in Biology*, Theoretical Biology, https://doi.org/10.1007/978-981-96-6026-1_7

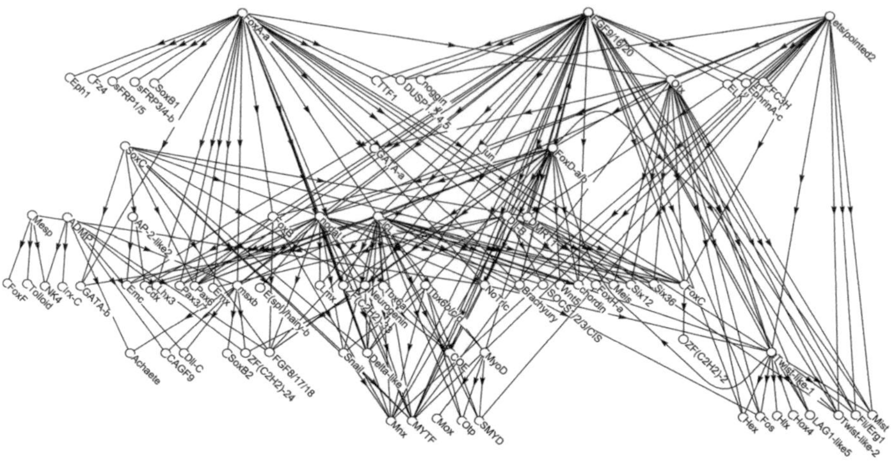

Fig. 7.1 Gene regulatory network containing 81 factors, responsible for cell-fate specification in the early development of ascidians. (Based on data from Imai et al. 2006; prepared by Mochizuki et al. 2013)

expression patterns in the embryo increases as different genes are activated in different cells. In addition, the gene expression pattern at a certain stage of development determines the fates of the cells, i.e., which tissue they will become in the future. In other words, the diversity of cell types in an organism's body is thought to derive from the diversity of gene expression patterns at a certain developmental stage.

How should we think about the fact that multiple gene expression patterns can arise from a single network system? Here, the behavior of a dynamical system after a sufficiently long time, or its asymptotic behavior, can help us understand it (see Mathematical Method 4A and Fig. 7.3 in Chap. 6). The set of points reaching with time evolution, starting from some initial value $\boldsymbol{u}_0$, is called the ω-limit set of $\boldsymbol{u}_0$. It is generally known that a non-linear dynamical system may have multiple ω-limit sets, depending on the initial conditions. It is thought that the emergence of multiple gene expression patterns from a single system during development is caused by multiple ω-limit sets in the corresponding dynamical system. The differences in the expression patterns between different tissues (or cells) may arise from different initial conditions caused by the cell-cell interactions or the biased cytoplasmic factors before the cell division. In the case of ascidian development, gene expression dynamics based on the network shown in Fig. 7.1 is considered to produce seven different expression patterns, which in turn produces seven different tissues. In other words, we can expect that the dynamics based on this gene regulatory network has the property of producing at least seven types of stable equilibrium points or periodic solutions.

Let us reconsider what is the information on the regulatory network from dynamical point of views. A regulatory network represents the information about "which gene influence which gene," that is, dependencies between variables in the network system. When we consider an ordinary differential equation (ODE) system

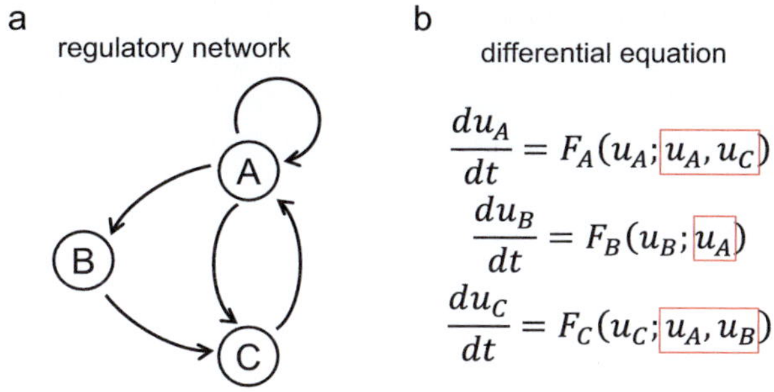

Fig. 7.2 The relationship between a regulatory network and differential equations. (Modified from Kobayashi et al. 2018). See Appendix for the meaning of the red boxes

corresponding to the focal system, the network provides information on the argument set of the right-hand-side functions of the ODE system (Fig. 7.2). On the other hand, since the network does not contain information on the mathematical formulation of the regulatory functions or parameter values in the functions, it is impossible to specify the details of the dynamics explicitly from the information of network, alone. For studying dynamics of network systems, many mathematical studies have taken a modeling approach where mathematical models are constructed by assuming functional forms and parameter values and trying to reproduce and explain observed biological phenomena.

However, in order to construct a mathematical model for a complex system like the one shown in Fig. 7.1, it is necessary to assume a very large number of functions and determine a huge number of parameter values. Even finding appropriate values requires a great deal of effort, and the obtained results will contain countless assumptions. It is clear to everyone that such a mathematical model is not useful. The difficulty in understanding the behavior of biological networks may be said to lie not in their complexity but in the lack of information for determining their dynamics.

In the next three chapters, we take a different approach from the standard way of modeling, namely model-free approach. We study an aspect (but an important aspect) of dynamical behaviors determined from network information alone. When we carefully consider what is important in life sciences, we may not necessarily need to determine all the details of dynamics. If we obtain a logical conclusion from network information alone without introducing other assumptions, it will be important and useful not only for the mathematical sciences, but also for the life sciences.

7.2 Structural Theory for Regulatory Networks: Linkage Logic

Consider dynamics on a regulation network $\Gamma = (V, E)$, where V is a set of vertices, and E is a set of directed edges (Mochizuki 2008; Mochizuki and Saito 2010; Fiedler et al. 2013; Mochizuki et al. 2013). Each vertex is assigned a quantity u_n ($n \in V$), and this quantity evolves over time according to the dependency relationships indicated by the regulatory network. Since the activity of biomolecules is the concentration of substances such as mRNA or proteins, these dynamics should include degradation or decay of activity. For example, consider a system of ordinary differential equations

$$\frac{du_n}{dt} = f_n(\boldsymbol{u}_{I_n}) - d_n(u_n), \quad n = 1, \cdots, |V| \tag{7.1}$$

where u_n is a positive value indicating activity or concentration of a biomolecule n. f_n is a positive function that indicates the activation of the molecule n, and is called the regulation function. Here, I_n denotes the argument set of f_n, indicating the factors that influence the activity of molecule n, i.e., $I_n := \{i | (i \to n) \in E\}$. $\boldsymbol{u}_{I_n}$ is a vector whose elements are u_i $(i \in I_n)$. d_n is a positive increasing function that represent decay or degradation of biomolecule n. In many biological systems, the specific forms of f_n and d_n are unknown. I_n contains n itself ($n \in I_n$), if and only if there is self-"activation" regulation. That is, the self-inhibition effect of u_n can be included in the decay term d_n, and need not be represented separately. Among self-regulations, only the self-activation, that cannot be represented by d_n, is included in f_n by setting $n \in I_n$. In the following, we consider regulatory networks redefined by I_n $(n \in V)$ constructed in this way. In other words, any self-regulatory arrows in the network below implies self-activation.

The form of Eq. (7.1) with the separate decay term was introduced for easier understanding. This form is not actually necessary, and the theory described here can be applied to more general formula of ODEs. Details are shown in Mathematical Method 7A.

Here we introduce an important concept in graph theory called a "feedback vertex set (FVS)" (Festa et al. 2000; Akutsu et al. 1998). An FVS is "a subset of the vertices in a directed graph such that, when it is removed, there are no directed cycles in the remaining graph." A directed cycle is a path in a graph that starts from a vertex, follows edges in the direction of the arrow, and returns to the original vertex.

Figure 7.3 shows examples of simple networks and their FVS. The top-left graph is a simple directed cycle containing three genes, and any one of these three constitutes the minimum FVS, since removing any one would eliminate the cycle. In the top-center graph, two vertices at the top and bottom can be excluded since the cycles only pass through the three central vertices. Removing only one from the central three vertices still leaves a cycle. Removing any two of the three vertices eliminates all cycles, so any two of the three vertices constitute the minimum FVSs. In the upper-right graph, any vertex with self-regulation ($n \in I_n$) is always contained in the FVS. In the bottom-left graphs, three loops pass through the central vertex. The bottom-right graph looks more complicated, but since all loops pass through the bottom right (or top right) vertex, the minimum FVS contains only one vertex in the bottom right (or top right). Choice of FVS and minimum FVS may not be unique.

FVS is a concept in graph theory. Mochizuki and Fiedler et al. found that this definition also plays an important role in the dynamics on graphs (Fiedler et al. 2013; Mochizuki et al. 2013). First, (1) *even if a system has multiple asymptotic behaviors,*

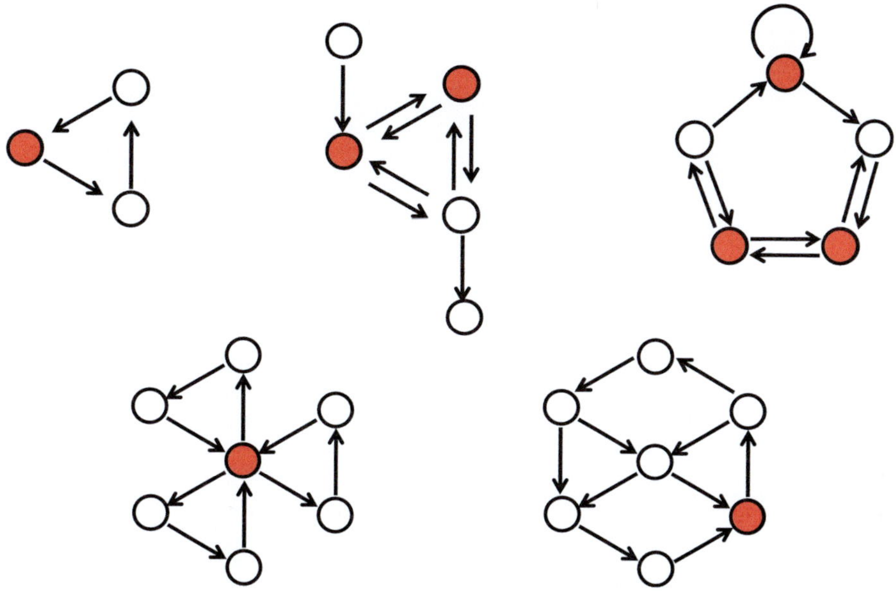

Fig. 7.3 Examples of hypothetical regulatory networks and minimum FVS. FVS (and minimum FVS) is not always unique. An example is colored here (Exercise 7.1)

the diversity of behaviors of the whole system should be detectable just by observing dynamics of FVS only. Second, (2) *by controlling behaviors of FVS to converge to one of any asymptotic trajectories, the whole system spontaneously comes to converge to the target trajectory*. These claims hold no matter what the nonlinear function of each vertex is. Conversely, the set of vertices that realize (1) observation and (2) control of the whole system without depending on model (functions and parameters) is only FVS. In other words, given information of network without knowing the functions, the minimum sufficient set of vertices to observe or to manipulate the whole system can be determined as minimum FVS.

The proof is given in the Mathematical Method 7A. Here, I give an intuitive explanation. First, consider a network with only one layer consisting of regulators (upper-side) and regulated (lower-side) node (Fig. 7.4a). Let us assume that the behaviors of the upper side are observable, i.e., the diversity of the behaviors of the upper side can be captured.

We focus on one of the observed upper-side behaviors. Given a behavior of the upper side uniquely, the behavior of the lower side should asymptotically approach a unique behavior after sufficient time (unless the lower side has self-loop, i.e., self-activation). Note that the unique determination does not imply determining the dynamics of the lower side "constructively," i.e., exact values of the lower side cannot be determined without knowing the functions f_n (or F_n in (7A.1)). However, we do not care about the details of the asymptotic behavior of the lower side. For the purpose of capturing the diversity of behaviors, it is not needed to observe the lower

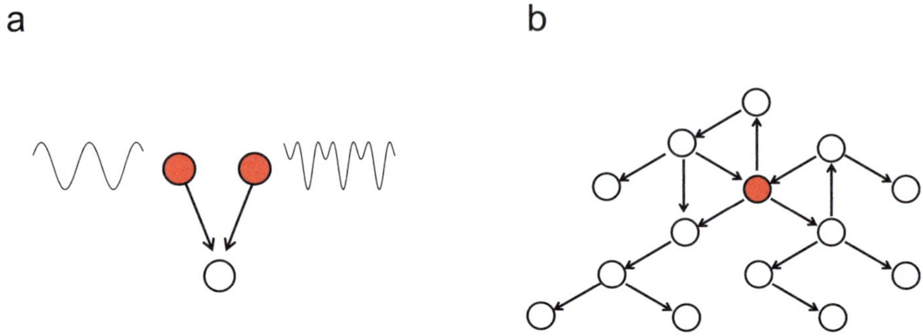

Fig. 7.4 An intuitive explanation of linkage logic theory. (**a**) Given the behavior of the upper vertices, the behavior of the lower vertex is determined uniquely. (**b**) Similarly, by giving (observing/manipulating) dynamics of an *appropriate* subset of vertices, the dynamics of the remaining vertices are determined uniquely. To minimize the set of vertices to observe/manipulate, under the constraint that dynamics of the remaining are determined uniquely, we should choose the minimum FVS

behaviors, because the solution of lower side should be unique for the given upper behavior.

Next, let us consider a network containing a large number of vertices (Fig. 7.4b). Even if the structure of the network is complicated, by choosing a set of vertices to observe "appropriately," the behavior of the remaining vertices is uniquely determined after a long time, and there is no need to observe them. Of course, the validity of this argument depends on how we choose the set of vertices to observe. Then, let us consider how we can minimize the set of vertices to be observed, while satisfying the condition that "the behavior of the remaining vertices is uniquely determined"? The answer is to choose the minimum FVS. From the definition of FVS, there are no directed cycles in the remaining network excluding FVS. Therefore, it should be possible to arrange all the remaining vertices in one direction from top to bottom, and their behavior should be uniquely determined in order from top to bottom. Therefore, it is sufficient to observe FVS only in order to identify all the behaviors of the entire network.

By simply replacing "observation" in the above argument with "manipulation," the controllability of the entire system by FVS is also shown in the same way. The underlying logic here is simply that, given the upper-side behavior, the unique determination of the asymptotic behavior of the lower side is guaranteed independent of the choice of functions. This logic is universal beyond the ODE-based formalism of dynamics. In fact, the importance of FVS in network dynamics was first pointed out in Boolean networks, which are discrete-time and discrete-state systems, by Akutsu et al. (1998).

A few additional explanations are in order. (i) For a given graph structure, FVS is the minimum target of observation/manipulation that guarantees observation/control of the whole system independent of the choice of functions. Depending on the choice of function, it may be possible to observe/control the whole system by a smaller set of variables than FVS. In other words, given only network information, observation/manipulation of the FVS is a "necessary and sufficient" condition for observing/controlling the whole dynamics, and for network systems with functions known, observation/manipulation of the FVS is a "sufficient" condition for observing/controlling the whole dynamics. (ii) A vertex with self-activation is always included in the FVS. In the logic shown in Fig. 7.4a, if the lower vertex has self-activation, its long-term behavior is not uniquely determined only from the upper vertex. However, this is not a problem, since this vertex has a self-activation, it is always included in the set to observe/manipulate, i.e., FVS. (iii) If there are vertices that do not receive any regulatory edge, they are not included in the minimum FVS. Considering the meaning of Eq. (7.1) (or (7A.1), (7A.2), it is obvious that the dynamics of such vertices converge uniquely to an equilibrium point, and do not contribute to the diversity of behaviors.

7.3 Control of Network Systems Based on Structural Theory

We show a numerical demonstration of this theory in this section (Mochizuki et al. 2013). Figure 7.5a shows the regulatory network for the human circadian rhythm. From the network, oscillation of gene expression with 24-hour cycle appears, which is the origin of the physiological oscillations in humans. Figure 7.5a shows an experimentally determined network consisting of 21 biomolecules, and its minimum FVS contains 7 molecules. As well as experimental studies, there have also been many mathematical studies on human circadian rhythms. In one of such mathematical studies, a 24-hour oscillation is reproduced in numerical simulation of a mathematical model, where the network in Fig. 7.5a is used. Hill functions, or mass action kinetics, are used for regulatory functions, and parameter values are chosen appropriately to produce the 24-hour oscillation (Mirsky et al. 2009).

We changed the parameter values slightly from the original study (Mirsky et al. 2009) to obtain behaviors different from that of the real circadian clock. Under a new parameter, the system has four invariant sets (Fig. 7.5b, c). Let us call the two stable periodic oscillations P1 and P2, the unstable periodic oscillation UP, and the unstable equilibrium point USS. In the following, we consider the control of switching dynamics between these four solutions. The procedure is as follows: (1) First, numerical simulations are performed, and the behaviors of FVS on the four solutions are recorded as a time series. (2) The dynamics are calculated numerically according to the ordinary differential equations of the model as the unmanipulated state. Then, the manipulation of FVS is started at a certain time point, where the behaviors of vertices in FVS are brought closer to the target behavior (one of the time series prepared in (1)). (3) The dynamics of variables other than the FVS factors are not manipulated and calculated following the ordinary differential equations as is. We observe how the behavior of these non-FVS variables changes as a result of manipulating FVS.

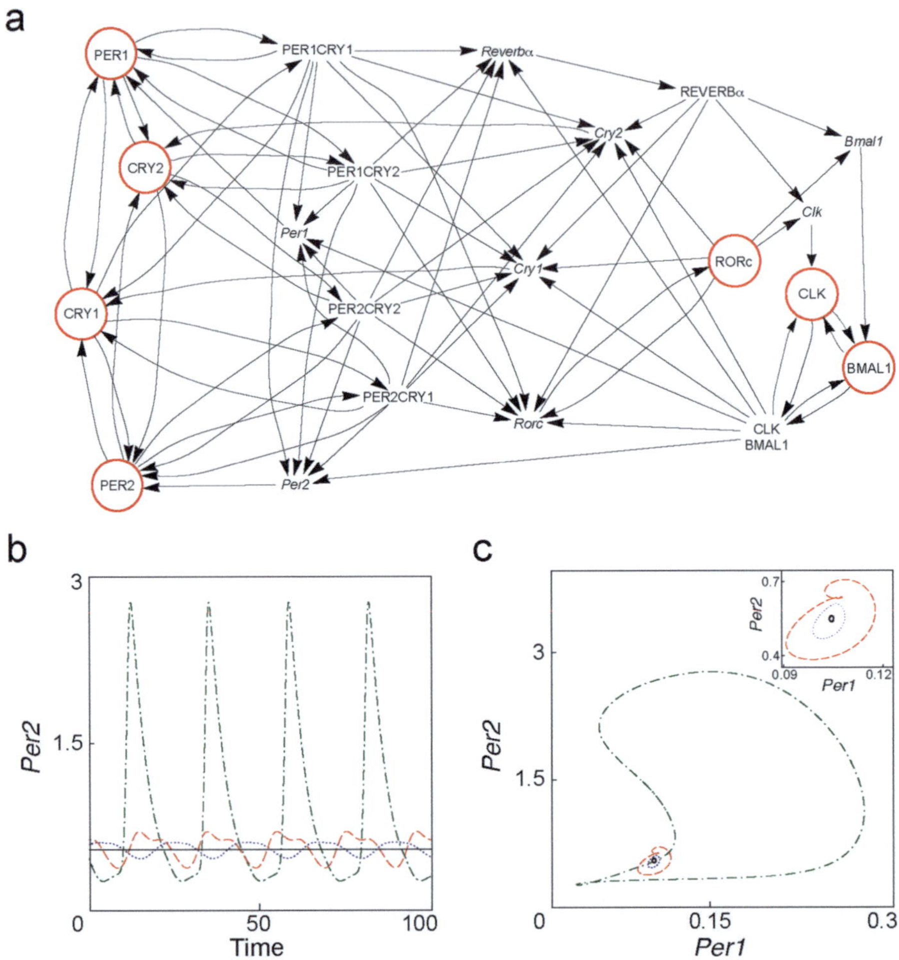

Fig. 7.5 Demonstration of controlling dynamics of gene regulatory network by FVS control. (**a**) The gene network that governs the human circadian rhythm. The red circle indicates one of the minimum FVS. (**b**, **c**) Four invariant sets of the dynamics: Green: P1, Blue: P2 (stable periodic oscillations), Red: UP (unstable periodic oscillations), Black: USS (unstable stationary state)

Figure 7.6 shows the behavior of two variables (*Per1*, *Per2*) not included in the FVS, obtained by the above manipulations. In Fig. 7.6a, the dynamics were on P1 before the manipulation. After starting manipulation of FVS to match the behavior on P2, the dynamics of the whole system quickly leave P1, approach P2, and then oscillate periodically on P2. Figure 7.6b shows the opposite, where the dynamics were oscillating on P2 before the manipulation. After manipulating the FVS to match the behavior on P1, the dynamics of the whole system converge to P1. In conclusion, controlling the whole system to converge to the target trajectory was achieved by manipulating FVS only.

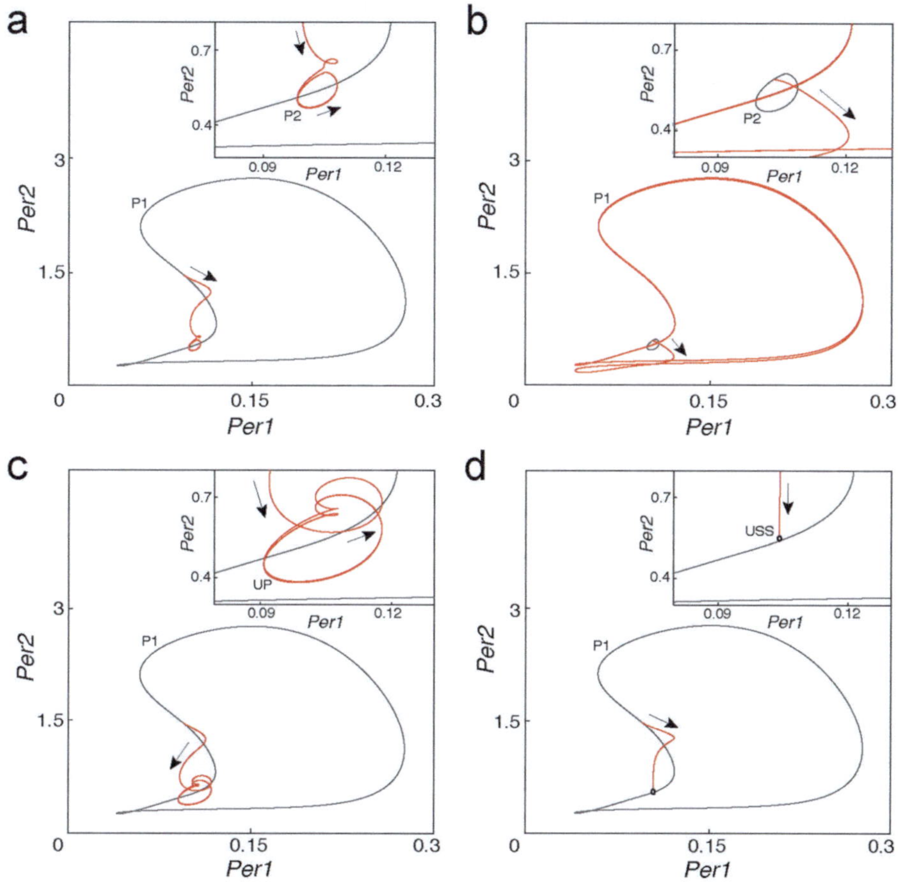

Fig. 7.6 Controlling dynamics of circadian rhythm by seven FVS factors. Dynamical behaviors on the plane (*Per1*, *Per2*), where *Per1* and *Per2* are not included in the FVS, are plotted on the plane. (**a**) From P1 to P2. (**b**) From P2 to P1. (**c**) From P1 to UP. (**d**) From P1 to USS

Figure 7.6c, d are particularly interesting. In Fig. 7.6c, the dynamics were on P1 before the manipulation. Then, FVS was manipulated to match the behavior of the unstable periodic oscillation UP. As a result, the whole system approaches and converges on this trajectory as if UP was a stable periodic solution. In Fig. 7.6d, for the dynamics on P1, FVS was manipulated and fixed to the value of the unstable equilibrium point USS. As a result, the whole system approaches and converges to this point, as if USS was the stable equilibrium point. In the original 21-variable system, UP and USS were of course unstable invariant sets. However, under the FVS manipulation, the dynamics show behaviors as if they were stable. The proof of the theorem shown in Mathematical Method 7A implies that under FVS manipulation, the dynamics always asymptotically approach a unique behavior from any initial state as if the system has a unique global solution (Fig. 7.7).

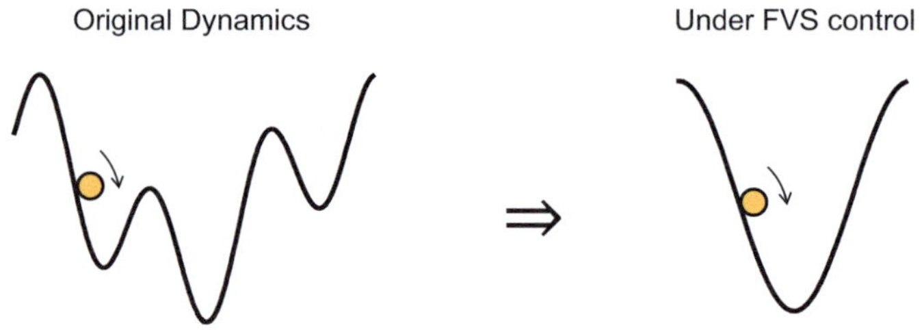

Fig. 7.7 Intuitive explanation for the FVS control. Even if the original dynamics have multiple stable solutions, under the FVS control the dynamics have a unique globally stable solution

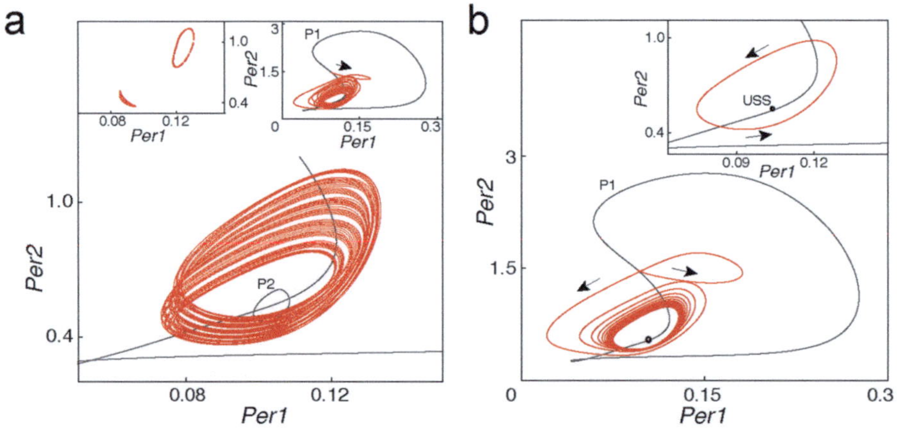

Fig. 7.8 Manipulation of a subset of FVS (including the six and excluding CLK) does not guarantee control

Figure 7.8 shows a numerical example showing that manipulation of an incomplete FVS does not guarantee the control of the whole system. The manipulation of the six variables in a subset of the FVS, excluding CLK, fails to lead the dynamics of the system to the unstable periodic solution UP or to the unstable equilibrium point USS. The dynamics converge to an unknown torus and oscillation solution, respectively.

In this case, the lack of one variable made it impossible to control the entire system, but as mentioned above, it is not always necessary to manipulate all the variables in the FVS to control the entire system. Depending on the choice of function for each vertex, it may be possible to observe/control the system with a smaller set of vertices than FVS.

7.4 Control of the Ascidian Gene Regulatory Network

In this section, we will demonstrate the utility of the theory by applying it to an actual biological network. As introduced at the beginning (Fig. 7.1), by the series of perturbation experiments, Sato's group determined the gene regulatory network, including 81 factors, from which 7 different cell types are generated during the early development of ascidians (Imai et al. 2006). Here, we attempt to understand the dynamics of cell fate specification in connection with the structure of this network (Fig. 7.9).

We first simplify the network: we repeatedly remove the genes that do not regulate anything and the genes that are not regulated by anything from the network in Fig. 7.1. Note that this operation does not affect the minimum FVS of the network. This operation results in a simple structure as shown in Fig. 7.9. The minimum FVS of the network contains only one vertex, Fox-Da/b. In other words, any asymptotic behaviors of the dynamics based on the network in Fig. 7.1 should be observed as the behavior of this gene. While this is an interesting result from a mathematical point of view, its implications for developmental biology are even more significant. In developmental biology, gene activity is often interpreted in binary terms: active or inactive. The ascidian, in particular, is an organism that exhibits a binary-like behavior of gene expression, with a clear switch between very high and almost zero activity of genes during the normal development. However, under a binary interpretation, the result of Fig. 7.9 implies that this gene network can produce only two states at the maximum.

There are two possible biological explanations for the result. The first explanation is that we should give up using the binary interpretation at this point to understand the dynamics of gene regulatory network of cell-fate specification of ascidian. If we take this explanation, we should measure the activity of a single factor, Fox-Da/b, in the FVS in more detail as a continuous quantity. Then, this single factor might exhibit seven different levels of gene expression as continuous quantities, which might reflect the seven different tissues. The second possible explanation is simple and radical: the network information may be incomplete and insufficient to

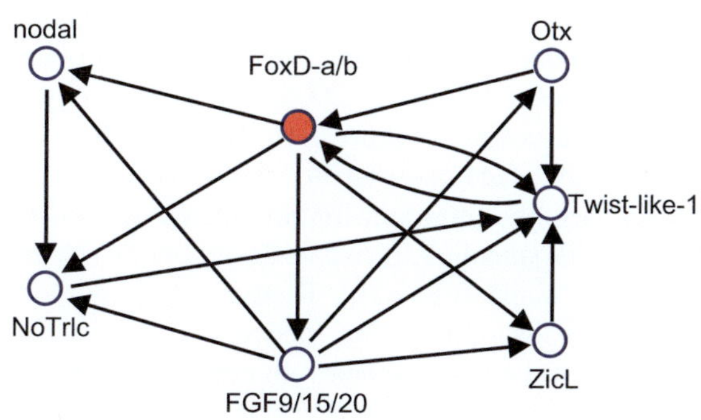

Fig. 7.9 Simplified gene regulatory network for ascidian early development. Upstream and downstream genes are removed from the network shown in Fig. 7.1, which is based on Imai et al. (2006). The minimum FVS is shown in red (Exercise 7.2)

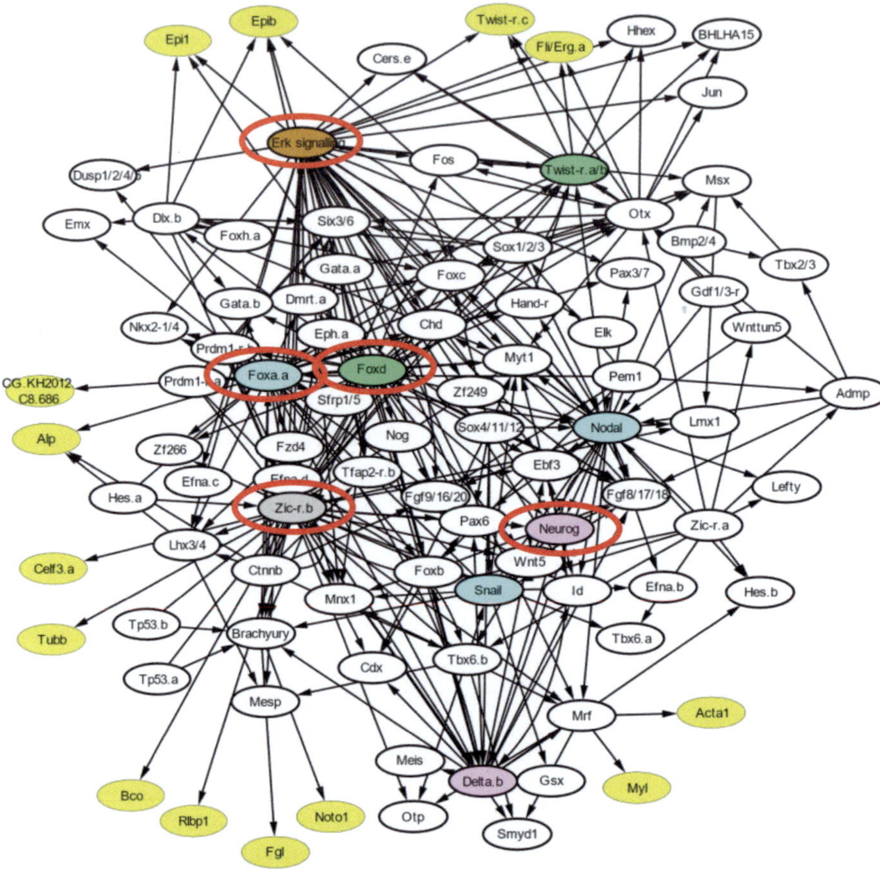

Fig. 7.10 Gene regulatory network for cell fate specification in ascidian early development (2015). This network contains 92 factors. The vertices circled in red indicate the selection of FVS factors by Kobayashi et al. (2018). Colored vertices mean that the FVS factors can be alternated between the same color. Vertices in yellow indicate marker genes for cell fates

understand diversity of cell fates in ascidian development. If we take this explanation, we have to reexamine the network and search for undiscovered regulations.

In fact, the project by the Sato's group continued after the regulatory network in Fig. 7.1 was published in 2006. By 2015, new gene regulations had been experimentally discovered. Figure 7.10 shows the 2015 version of the regulatory network including 92 factors. Analysis of this network revealed that, despite its apparent complexity, the minimum FVS contains only five vertices [10]. The 5 factors are Zic-r.b, Erk signaling, (Foxd or Twist-like1), (Neurogenin or Delta.b), and (Foxa.a, Nodal, or Snail) (Fig. 7.10). There are 12 possible ways to choose the set.

The results of the 2006 network analysis seemed insufficient to explain the diversity of the seven species. In contrast, the results of the 2015 network analysis give us an important message that is of great interest to biology. In other words, if the

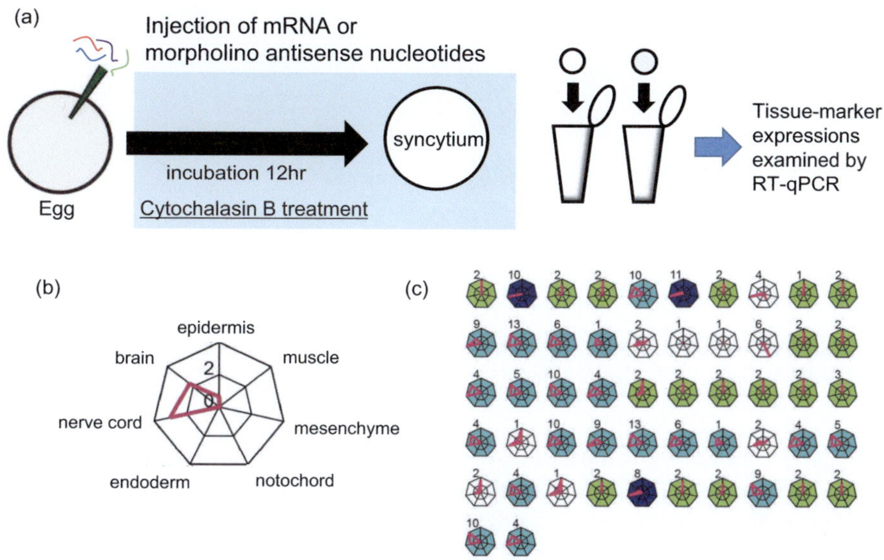

Fig. 7.11 Experiments of controlling FVS of gene regulatory network of ascidians. (**a**) Manipulation of factors in the FVS by introducing nucleotides into fertilized eggs or by adding chemicals to the culture medium. After culture, the cell fate is identified by the expression of marker genes. (**b**) The population average of marker gene expression under "no manipulation" of FVS factors. (**c**) Marker gene expression of each individual experiment under no manipulation. The numbers on the radar chart indicate how many times larger the maximum of the chart is compared to the expression level of normal embryos. In (**c**), the background color of the radar chart indicates the observed tissue-specific expression. Green: epidermis, light blue: brain, dark blue: nerve cord

information in the network is complete (or sufficient) and fate determination is indeed based on the dynamics of this network, then it should be possible to arbitrarily reproduce cell fates in the seven tissues by manipulating these five genes. To verify this prediction, Kobayashi et al. conducted an experiment to manipulate the five factors (Kobayashi et al. 2018).

From 12 minimum FVSs, we selected the one that was easy to manipulate experimentally (Foxa.a, Foxd, Neurogenin, Zic-r.b, Erk). Note that these genes (except for Erk signaling) code transcription factors. Each of these factors is known to be active in multiple different lineages in ascidian development and is not specific to unique cell type. Next, in order to exclude the effects of cell-cell interactions, we created a single-cell embryo system by cytochalasin B treatment. Under the treatment, cytokinesis was halted while expression dynamics proceeded (Fig. 7.11). The activities of Foxa.a, Foxd, Neurogenin, and Zic-r.b were manipulated by introducing mRNA (activator) or morpholino antisense oligonucleotides (inhibitor) into the fertilized eggs, and Erk signal was manipulated by adding FGF or U0126 to the culture medium. After the manipulated embryos were cultured and grown for a while, the expression of marker genes corresponding to the seven tissues (epidermis, brain, nerve cord, muscle, notochord, mesenchyme, and endoderm) was measured, and the results of manipulation were examined.

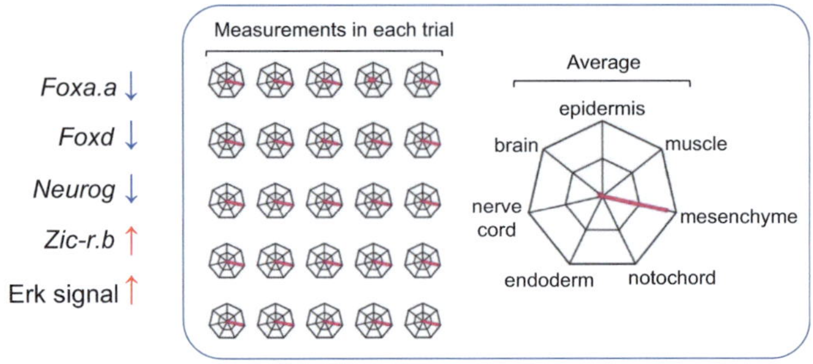

Fig. 7.12 An example of results of FVS control. Under this manipulation, only a single marker corresponding to the mesenchyme tissue was always active without variety between individuals

Figures 7.11b, c show the expression of marker genes without any manipulation of the FVS factors as a control experiment. It turns out that only a limited cell fates, i.e., epidermis, brain, or nerve cord were induced in the single-cell embryo without FVS control. The fates of embryos are not uniquely determined, as marker gene expressions were varied between embryos.

Figure 7.12 shows an example of results of manipulation experiments, where three factors were suppressed and two factors were activated. Foxa.a, Foxd, and Neurogenin were suppressed using morpholino antisense oligos, Zic-r.b was activated by introducing the mRNA, and the Erk signaling pathway was activated by adding FGF to the culture medium. The experiment was repeated 25 times using different embryos, and in all cases (except for one case in which no marker genes were expressed), it was observed that only the mesenchyme markers were activated. In other words, under this manipulation, the fate of the embryo is deterministically directed to the mesenchyme uniquely.

Under the binary manipulation, each of the five FVS factors is either activated or suppressed, i.e., there are $32(=2^5)$ possible manipulations. A total of 32 exhaustive manipulation experiments were performed to capture all possible states that this system can produce. The results are summarized in Fig. 7.13. In many cases of the exhaustive manipulations, the resultant marker gene expression shows the deterministic induction of a unique cell fate, as observed in the example shown in Fig. 7.12. For example, when all five FVS factors were inactivated, only epidermis markers were always expressed, and when only Foxa.a was activated and the remaining four were inactivated, only endoderm markers were always expressed. When we read Fig. 7.13, it is important to note that not only the activated factors are important for the resultant induction, but it is also important to manipulate each of all factors in the FVS, including the inactivated factors.

Throughout the exhaustive manipulations, there are methods that deterministically and uniquely induce the marker expression for each of the epidermis, endoderm, notochord, nerve cord, brain, and mesenchyme. In other words, six out of the seven tissues observed in normal development, except for muscle, could be induced by the FVS control (Kobayashi et al. 2018). In developmental biology, various

Foxa.a	*Foxd*	*Neurog*	*Zic-r.b*	Erk signal	marker expression	*Foxa.a*	*Foxd*	*Neurog*	*Zic-r.b*	Erk signal	marker expression
Down	Down	Down	Down	Down	epidermis	Up	Up	Down	Down	Up	none
Down	Down	Down	Down	Up	none	Up	Down	Up	Down	Up	nerve cord
Up	Down	Down	Down	Down	endoderm	Up	Down	Down	Up	Up	nerve cord
Down	Up	Down	Down	Down	notochord	Down	Up	Up	Down	Up	nerve cord
Down	Down	Up	Down	Down	nerve cord	Down	Up	Down	Up	Up	none
Down	Down	Down	Up	Down	brain	Down	Down	Up	Up	Up	nerve cord
Up	Down	Down	Down	Up	none	Up	Up	Up	Down	Down	notochord
Down	Up	Down	Down	Up	none	Up	Up	Down	Up	Down	endoderm
Down	Down	Up	Down	Up	verve cord	Up	Down	Up	Up	Down	nerve cord
Down	Down	Down	Up	Up	mesenchyme	Down	Up	Up	Up	Down	endoderm
Up	Up	Down	Down	Down	notochord	Up	Up	Up	Down	Up	notochord
Up	Down	Up	Down	Down	nerve cord	Up	Up	Down	Up	Up	endoderm
Up	Down	Down	Up	Down	nerve cord	Up	Down	Up	Up	Up	brain
Down	Up	Up	Down	Down	none	Down	Up	Up	Up	Up	none
Down	Up	Down	Up	Down	none	Up	Up	Up	Up	Down	none
Down	Down	Up	Up	Down	brain	Up	Up	Up	Up	Up	none

Fig. 7.13 Marker gene expression in the exhaustive $2^5 = 32$ manipulations of FVS of the gene regulatory network in ascidians. Marker genes specifically expressed in each FVS-manipulated embryo are shown. "None" indicates that there is no predominantly expressed gene

experimental methods have been developed to induce tissues, but all of these methods are empirical and specific to a single target tissue. There has never been an achievement in which the manipulation criteria were proposed based on mathematical theory, and various cell fates were systematically induced within the same framework.

At the same time, however, it became clear that muscle cells could not be induced, even though the FVS was manipulated exhaustively (albeit within a binary range). According to the mathematical theory, an FVS is the minimum and sufficient manipulation target to reproduce all of the behaviors of the network system. The fact that all of the expected behaviors could not be reproduced despite the exhaustive manipulation may suggest that there is still some missing information in the network. Another possibility, of course, is that cell-cell interactions other than the regulatory network play an important role in determining muscle fate and cannot be understood in terms of dynamics within a single cell. However, the ascidian muscle tissue is a well-known example of autonomous development for more than 100 years; cell-cell interactions are unlikely to play essential roles in this regard.

There are cases where the addition of a single regulatory edge to the regulatory network changes the FVS. These may be candidates for unknown regulation. Therefore, a list of candidate unknown regulations that could change FVS by hypothetically adding to the regulatory network was created. The list was examined by the Sato lab, and it was confirmed that one of the regulations in the list, from Mrf to Tbx6-r.b, actually exists and works in the development of ascidian. Adding this new regulation to the network, a minimum FVS of size 6 was obtained including Tbx6-r.b, {Foxa.a, Foxd, Neurogenin, Zic-r.b, Erk, Tbx6-r.b}. In other words, the reason why all the tissues could not be reproduced might be because only five factors, excluding Tbx6-r.b, were manipulated in the previous experiment, whereas actually these six factors should have been manipulated.

However, to perform an exhaustive manipulation of the six factors in the new FVS, $64(=2^6)$ manipulation experiments were required, making the project even

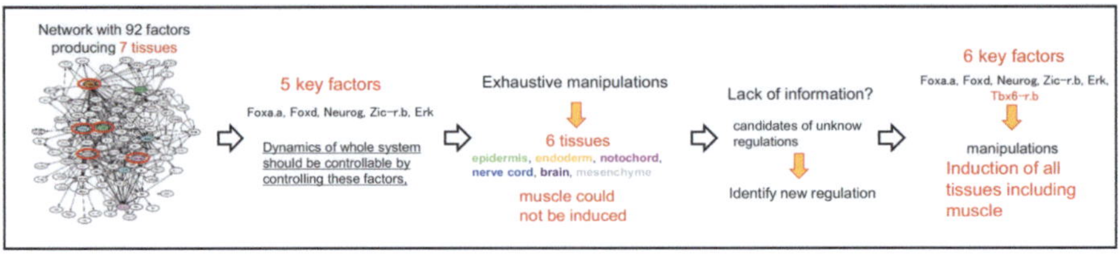

Fig. 7.14 Summary of the analysis of the ascidian gene regulatory network using linkage logic and experimental validation

more difficult than the 32 manipulation experiments. Therefore, an experiment was conducted on a trial basis in which the newly added factor, Tbx6-r.b, was activated and the five factors originally included in FVS were deactivated. Although this was only a pilot experiment, surprisingly, only the muscle markers were induced. Furthermore, it was also confirmed that the six other tissues that were known to be induced by manipulation of the five factors were also induced by manipulation of the six FVS factors. This showed the insufficiency of the five-factor manipulation and the sufficiency of the six-factor manipulation (Kobayashi et al. 2021) (Fig. 7.14).

Let us summarize the results of this section. By applying linkage logic theory to the ascidian gene regulatory network, it was demonstrated that the cell fate-specification system can indeed be controlled by manipulating the activity of the determined molecules. It is a large achievement to determine the important molecules from the network structure alone, without using information on the biological properties of the molecules, and to succeed in complete control of the complex system. In addition, it is notable that the network, which serves as the basic information, was updated through the exchange of mathematical prediction and experimental test. We would say that this was achieved since linkage logic is a model-free theory without introducing assumptions of functions or parameter values. It is a fairly strong proposition from the theoretical side that "the expected results were not obtained because the data were incomplete." This argument was possible because the results of theoretical analysis do not depend on model assumptions. Experimental verification of the prediction from linkage logic is not just confirmation but has the meaning of testing the completeness of the network information.

7.5 Summary

The linkage logic presented in this chapter is a mathematical theory that directly connects the network structure and dynamics of regulatory systems. The conclusion from the theory can be summarized quite simply: To understand the dynamics of complex regulatory networks, we should focus on the FVS. This is because the FVS is the minimum sufficient subset of variables to capture the dynamics of the whole system, and at the same time, the minimum sufficient subset of variables to control the whole system to converge to any desired behavior.

The linkage logic theory introduced here is not only valid for specific system of ascidian cell-specification but can also be applied to any regulatory networks. Once we have enough information on regulatory relationships between biomolecules, we can discuss the behavior of the system of any organisms. Moreover, the mathematical formulation of our definition of a regulatory system implies that this theory is applicable not only to biological regulatory networks but also to any regulatory relationships. We may expect that the theory, which originally developed for living systems, will be used for understanding behaviors of high-dimensional dynamical systems, like economic systems or human relationships networks.

Mathematical Method 7A (Structural Theory for Regulation Networks)

1. Formulation

Consider a directed graph $\Gamma = (V, E)$ where V is a set of vertices, and E is a set of directed edges, where vertices represent biomolecules and edges represent regulatory linkages. Suppose that the dynamics of molecular activity on this graph is described by a system of ordinary differential equations in the form (Fiedler et al. 2013; Mochizuki et al. 2013):

$$\begin{aligned}\frac{du_n}{dt} &= F_n(\boldsymbol{u}) \\ &= F_n(u_n, \boldsymbol{u}_{I_n})\end{aligned}, \qquad n = 1, \cdots, |V| \tag{7A.1}$$

Here, $\boldsymbol{u}_{I_n}$ is a vector notation with u_i $(i \in I_n)$ as elements. Unlike f_n in (7.1), F_n can be either positive or negative.

We also assume that (i) each F_n is a continuous and differentiable function, (ii) there is no explosive solution (F_n is bounded), and (iii) F_n satisfies the "decay condition":

$$\partial_1 F_n(u_n, \boldsymbol{u}_{I_n}) < 0. \tag{7A.2}$$

Here $\partial_1 F$ has a special meaning: "the partial derivative of F with respect to the first argument." This Eq. (7A.2) corresponds to the decay term or self-inhibition in Eq. (7.1). The set $I_n \subseteq V$ is the input set of n, a subset of molecules that regulate molecule n, i.e. $I_n := \{i \mid (i \rightarrow n) \in E\}$. The set I_n includes n $(n \in I_n)$ i.e. a self-regulation, if and only if $\partial F_n / \partial u_n$ is "not always negative." Because the effects of decay or self-inhibition are already represented by the decay condition (7A.2). If there exist effects depending on u_n that cannot be expressed by the decay condition (7A.2), it is necessary to include n in I_n. In the discussion below, we consider that the sets of I_n ($\forall n \in V$) directly represent the graphical structure of the regulatory network, i.e. the regulatory network is defined by the set of I_n. That is, if there exists a self-regulatory loop in a graph, it indicates an autoactivation regulation.

Note that the decay condition $\partial_1 F_n < 0$ does not require the system always satisfy $\partial F_n/\partial u_n < 0$. Suppose that we have a differential equation where $\partial F_n/\partial u_n \geq 0$. In that case, using $\tilde{I}_n := I_n \cup \{n\}$ we can reconstruct the function as

$$\tilde{F}_n(\eta_n, \boldsymbol{u}_{\tilde{I}_n}) := F_n(\boldsymbol{u}_{I_n}) + u_n - \eta_n \tag{7A.3}$$

where the decay condition (7A.2) is satisfied. In other words, when there is no decay term in the mathematical formula, we can interpret that there is a self-activating term that cancels out the decay term, which is assumed to exist. Thus, the requirement of decay conditions does not constrain the mathematical formula of the differential equations.

2. Two Definitions

Definition 1 (Feedback Vertex Set: FVS)

In a directed graph $\Gamma = (V, E)$, a subset $I \subseteq V$ of vertices is called a feedback vertex set (FVS), if the graph $\Gamma \setminus I$ excluding the subset I has no directed cycles.

FVS is one of the major concepts in graph theory, and it has been shown that finding an FVS for any graph is NP-hard problem (Festa et al. 2000; Akutsu et al. 1998).

Definition 2 (Determining nodes)

In a dynamical system, a subset $J \subseteq V$ of variables is called a set of determining nodes, if the following condition is satisfied:
For any two solutions $\boldsymbol{u}(t)$ and $\tilde{\boldsymbol{u}}(t)$ of the dynamics, convergence $\tilde{\boldsymbol{u}}_J(t) - \boldsymbol{u}_J(t) \to 0$ ($t \to +\infty$) of $\boldsymbol{u}_J(t)$, $\tilde{\boldsymbol{u}}_J(t)$ on the subset J always implies the convergence $\tilde{\boldsymbol{u}}(t) - \boldsymbol{u}(t) \to 0$ ($t \to +\infty$) of two solutions $\boldsymbol{u}(t)$, $\tilde{\boldsymbol{u}}(t)$ of the whole system V.

$\boldsymbol{u}(t)$ and $\tilde{\boldsymbol{u}}(t)$ are "two solutions starting from different initial conditions." In the context for controllability, we interpret that $\tilde{\boldsymbol{u}}(t)$ is the target behavior and $\boldsymbol{u}(t)$ is the behavior of the system under control.

A set of determining nodes is a subset of variables such that one can identify any asymptotic behaviors of the whole system by observing them without observing all the variables in the system (Foias and Temam 1984). By considering the contrapositive of Definition 2, it is easy to understand that the set of determining nodes captures the diversity of the asymptotic behaviors of the dynamical system. The contrapositive ($\tilde{\boldsymbol{u}}(t) - \boldsymbol{u}(t) \nrightarrow 0 \Rightarrow \tilde{\boldsymbol{u}}_J(t) - \boldsymbol{u}_J(t) \nrightarrow 0$) means that "if two solutions do not asymptotically converge to the same behavior, then (at least one of) the variables

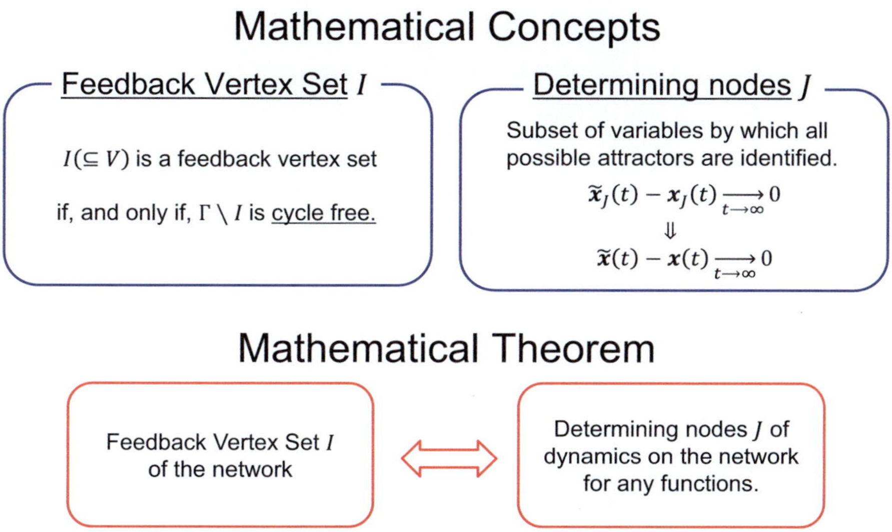

Fig. 7.A1 The structure of the theorem

in the subset J do not asymptotically converge to the same behavior." A concept of determining nodes has been discussed in the field of dynamical systems theory, especially in the context of the Navier-Stokes equation, and its relationship with graphs has not been discussed at all.

3. The Theorem and Its Proof

Theorem (FVS = set of determining nodes)

> In the dynamics (7A.1) on a directed graph, an FVS of the graph is a set of determining nodes regardless of the choice of F_n. Conversely, if a subset of variables is a set of determining nodes regardless of the choice of F_n, then it is an FVS of the graph (Fig. 7.A1).

Here is an outline of the proof. First, we show that "FVS $\Rightarrow$ Set of determining nodes."

(1) For a given FVS I, a set $K := V \setminus I$ of vertices not included in I is called the nonFVS. From the definition of FVS, a directed cycle cannot be composed only of the vertices in nonFVS K. Therefore, for each vertex $k \in K$ in the nonFVS, the number k can be reassigned so that $I_k \subseteq I \cup \{1, \ldots, k-1\}$ $(\forall k \in K)$. That is, $k \in K$ is reordered so that a regulator vertex has a younger number than the regulated vertex for each regulatory edge.

(2) Let $w_n(t) := \tilde{u}_n(t) - u_n(t)$ be the difference in variable u_n between two dynamical solutions $\boldsymbol{u}(t)$ and $\tilde{\boldsymbol{u}}(t)$ at time t. Suppose that $w_i(t) \to 0$ is given for all vertices $\forall i \in I$ in the FVS I, and show by mathematical induction that $w_k(t) \to 0$ for each vertex $k \in K$ in the nonFVS K. For the dynamics of the difference of solutions $\boldsymbol{w}(t) := \tilde{\boldsymbol{u}}(t) - \boldsymbol{u}(t)$, we can derive.

$$\begin{aligned}\frac{d}{dt}\boldsymbol{w}(t) &= \frac{d}{dt}\tilde{\boldsymbol{u}}(t) - \frac{d}{dt}\boldsymbol{u}(t)\\ &= \boldsymbol{F}(\boldsymbol{u}(t)+\boldsymbol{w}(t)) - \boldsymbol{F}(\boldsymbol{u}(t))\\ &= [\boldsymbol{F}(\boldsymbol{u}(t)+\theta\boldsymbol{w}(t))]_{\theta=0}^{1}\\ &= \int_0^1 \frac{d}{d\theta}\boldsymbol{F}(\boldsymbol{u}(t)+\theta\boldsymbol{w}(t))d\theta\\ &= \int_0^1 \frac{\partial}{\partial\boldsymbol{u}}\boldsymbol{F}(\boldsymbol{u}(t)+\theta\boldsymbol{w}(t))\cdot\boldsymbol{w}(t)d\theta\\ &= A(t)\boldsymbol{w}(t)\end{aligned} \tag{7A.4}$$

$$A(t) := \left(\int_0^1 \frac{\partial \boldsymbol{F}}{\partial \boldsymbol{u}}\bigg|_{\boldsymbol{u}(t)+\theta\boldsymbol{w}(t)} d\theta\right) \tag{7A.5}$$

where A is the matrix obtained by integrating each component of the Jacobian matrix of $\boldsymbol{F}$ at $\boldsymbol{u}(t) + \theta\boldsymbol{w}(t)$ with respect to θ. Note that any vertex in K does not have self-regulation, i.e., self-activation. In other words, $F_k/\partial u_k < 0\ \forall k \in K$. Then dynamics of each $k \in K$ can be represented by a linear nonautonomous dynamic equation,

$$\frac{d}{dt}w_k(t) = -d_k(t)w_k(t) + \sum\nolimits_{k' \in I_k} a_{k,k'}(t)w_{k'}(t), \tag{7A.6}$$

where the first term correspond to the decay condition (7A.2) and the second term is from the second argument set in (7A.1). The nonautonomous coefficients $d_k(t) \in \mathbb{R}$ and $a_{k,k'}(t) \in \mathbb{R}$ are bounded by positive constants d_0 and a_0 as $0 < d_0 \le d_k(t)$ and $|a_{k,k'}| \le a_0$, respectively. Assuming that $w_n(t) \to 0$ has already been shown for $\forall n \in \{1, \cdots, k-1\}$, let us show for $n = k$. Solving the equations yields:

$$\begin{aligned} w_k(t) &= \exp\left(-\int_0^t d_k(s)ds\right)w_k(0) + \sum_{k'\in I_k}\int_0^t \exp\left(-\int_s^t d_k(\sigma)d\sigma\right)a_{k,k'}(s)w_{k'}(s)ds\\ &\le \exp\left(-\int_0^t d_k(s)ds\right)|w_k(0)| + \sum_{k'\in I_k}\int_0^t exp\left(-\int_s^t d_k(\sigma)d\sigma\right)|a_{k,k'}(s)||w_{k'}(s)|ds\\ &\le \exp(-d_0t)|w_k(0)| + \sum_{k'\in I_k}\int_0^t \exp(-d_0(t-s))a_0|w_{k'}(s)|ds\\ &\to 0 \quad (t \to \infty)\end{aligned} \tag{7A.7}$$

Note that the convergence of the first term is shown by $\exp(-d_0t) \to 0$ $(t \to \infty)$, and the convergence of the second term is shown by $w_{k'}(s) \to 0$ $(s \to \infty)$ for $k' \in I_k \subseteq I \cup \{1, \ldots, k-1\}$. When $k = 1$, then $I_1 \subseteq I$, it is obvious that $w_1(t) \to 0$ from the similar argument. From the above, the first half of the theorem is proved.

"Set of determining nodes ⇒ FVS'" is shown by taking the contrapositive, that is, a subset of vertices which is not an FVS is not a set of determining nodes. Suppose I' is not an FVS, i.e., $\Gamma \setminus I'$ contains directed cycles. By choosing the function of the vertices appropriately, dynamics can be constructed so that I' is not a set of determining nodes. For example, all functions included in I' are taken to be simple decay $F_n(x_n, \boldsymbol{x}_{I_n}) := -x_n$. From this, the behavior of $\Gamma \setminus I'$ cannot be captured by I'. However, there is a cycle within $\Gamma \setminus I'$ and, by choosing these functions, it is possible to create diversity in the solutions such as multiple stationary points. In other words, I', which is not an FVS, is not always a set of determining nodes for arbitrary functions. (End of proof).

Here we add a supplementary note on the control of dynamic by manipulating the FVS. The theorem claims that when the behaviors $\boldsymbol{u}_{FVS}(t)$ of vertices in the FVS are made closer to the target behaviors $\tilde{\boldsymbol{u}}_{FVS}(t)$ ($\tilde{\boldsymbol{u}}_{FVS}(t) - \boldsymbol{u}_{FVS}(t) \to 0$ $(t \to +\infty)$), the behavior $\boldsymbol{u}(t)$ of the whole system spontaneously approaches the target behavior $\tilde{\boldsymbol{u}}(t)$ ($\tilde{\boldsymbol{u}}(t) - \boldsymbol{u}(t) \to 0$ $(t \to +\infty)$). The target $\tilde{\boldsymbol{u}}_{FVS}(t)$ does not have to be an ω-limit set such as stable equilibria or stable periodic oscillations, but can be an α-limit set such as unstable equilibrium points or unstable periodic solutions. Moreover, if we pay careful attention to the theorem, the target need not be the invariant set of the dynamical systems. Given any behaviors in the FVS, the entire system will asymptotically approach a uniquely determined behavior (whatever that may be). In other words, under any manipulation of FVS, the system loses the diversity of asymptotic behaviors. When FVS approaches the system's original asymptotic behavior, the entire system uniquely asymptotes to that behavior.

Mathematical Method 7B (Controlling Theory of Linear Network Systems)

In this chapter, we have introduced a method for determining the vertices to be manipulated from the network structure in order to control a nonlinear dynamical system and make it converge to an arbitrary behavior. On the other hand, in the field of control theory, another criterion for determining the driver nodes has been discussed for linear dynamical systems.

Consider the following linear dynamical system

$$\frac{d\boldsymbol{u}(t)}{dt} = A\boldsymbol{u}(t) + B\boldsymbol{v}(t) \tag{7B.1}$$

where $\boldsymbol{u}(t) \in \mathbb{R}^N$ is the state vector, $\boldsymbol{v}(t) \in \mathbb{R}^M$ is the input vector, A is an $N \times N$ matrix, and B is an $N \times M$ matrix. The regulatory structure of the system is represented by the components of the matrix A being either zero or nonzero.

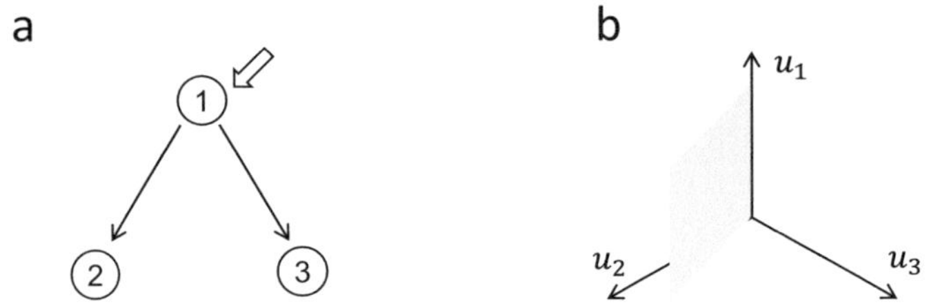

Fig. 7.B1 Example of a simple regulation system and controllable subspace

Kalman's controllability condition is given by the full-rankness of the controllability matrix (Kalman 1963), i.e.

$$\mathrm{rank}\left[B\,AB\,A^2B\cdots A^{N-1}B\right]=N \tag{7B.2}$$

Let us illustrate with a simple example. Consider the regulatory relationship among the three factors as shown in Fig. 7.B1a. Assuming that vertex 1 is the vertex to be manipulated, the matrices A and B in Eq. (7B.1) are given, respectively, as follows

$$A=\begin{pmatrix} 0 & 0 & 0 \\ a & 0 & 0 \\ b & 0 & 0 \end{pmatrix}, B=\begin{pmatrix} 1 \\ 0 \\ 0 \end{pmatrix} \tag{7B.3}$$

Then the controllability matrix is

$$\begin{pmatrix} 1 & 0 & 0 \\ 0 & a & 0 \\ 0 & b & 0 \end{pmatrix} \tag{7B.4}$$

The controllable subspace is shown in gray in Fig. 7.B1b. In this case, the position across the (u_2, u_3) plane changes depending on the ratio of a to b, but the controllable subspace is always two-dimensional. In this case, the controllable matrix is not full rank and is not controllable by Kalman's criteria.

Based on these criteria, Lin and Liu et al. provided guidelines for selecting the vertices to be manipulated (driver nodes), i.e., the nonzero components of B (and the number of columns M), from the network structure alone, i.e., the distribution of nonzero components of A (Lin 1974; Liu et al. 2011). Here, B is assumed to have only one nonzero component in each column. Let us call it the "maximum matching method" to distinguish it from the controlling method introduced in this chapter (hereafter referred to as "FVS control"). The guideline is given by the following algorithm (Fig. 7.B2).

1. For a given directed graph, determine the largest subset of edges that do not share starting points with each other and do not share ending points with each other (called "maximum matching" in graph theory).

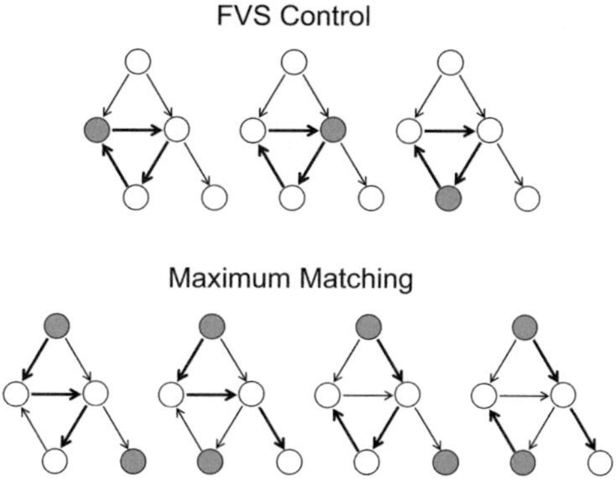

Fig. 7.B2 Comparison of the two methods of selecting driver nodes: FVS control selects the smallest vertex set that cuts all the cycles (the thick edges). Maximum matching selects the vertex set that is not the endpoint of the chosen edge in the maximum matching set (bold line)

2. The vertices that are the endpoints of the edges in this set are called "matched vertices", and the others are called unmatched vertices.
3. If there exists a directed path from the set of unmatched vertices to all vertices, then this unmatched vertex set is the smallest set of driver nodes that can satisfy the Kalman controllability condition.

A few additional remarks. The set of edges satisfying the maximum matching may not be unique. Therefore, the minimum set of driver nodes determined by this method may not be unique. The vertices that are not the endpoints of any edges (i.e., not regulated by any vertices) are always included in the set of driver nodes.

The method is based on the following understanding. Suppose a vertex is controllable in the Kalman's sense. If only one edge emanates from the vertex, the vertex at its tip is controllable, but if multiple edges emanate from the vertex, the vertices at the tips of the edges cannot be independently controllable. Based on this understanding, maximum matching means selecting only one edge from the edges emanating from each vertex and avoiding redundancy such that multiple selected edges point to the same vertex. This method determines the driver nodes without considering feedback of the graph. In other words, the selection of driver nodes by this method is usually different from minimum FVS. Figure 7.B2 compares the selection of driver nodes by the criteria of FVS control and the maximum matching method using a simple network.

Liu et al. calculated the size of the smallest set of driver nodes determined by this algorithm for various network systems, including biological and social. Zanudo et al. similarly compared the size of the set of driver nodes between the FVS control and the maximum matching method for various network systems (Zanudo et al. 2017). Which method is advantageous (providing smaller set) depends on the structure of the network. Although the proof of controllability by the maximum matching method is guaranteed only for linear systems, studies have attempted to control nonlinear systems numerically using the guidelines. There have been both positive and negative reports on the possibility of controlling nonlinear systems with this method.

In contrast, the FVS control introduced in this chapter guarantees controllability of the dynamics for any nonlinear system as long as differentiability and boundedness (dissipativity) are satisfied. Although the method of Lin and Liu et al. determines the vertices to be manipulated from the structure of the network, the input function $\boldsymbol{v}(t) \in \mathbb{R}^M$ is not given, and this must be determined specifically for each model. In particular, in the case of nonlinear systems, $\boldsymbol{v}(t)$ may need to be modulated with time, but no guidelines for determining $\boldsymbol{v}(t)$ are given in this case. In contrast, in FVS control, after identifying the FVSs that are subject to manipulation from the network, the method of operation is simply to make the behavior of FVSs match the target behavior. Not only in the mathematical theory, but also in the manipulation in practice, no information about the model is required at all.

These differences may be due to differences in the meaning of "control." The maximum matching method is based on Kalman's controllability condition and aims to guide the system in all directions in linear state space. FVS control, on the other hand, aims to guide the system toward one of its intrinsic asymptotic behaviors. In other words, FVS control is a method that maximizes the use of the system's intrinsic dynamical properties.

Exercise

Exercise 7.1

Find the minimum FVS for each of the five regulatory networks shown in Fig. 7.3 of this chapter. If there is more than one, list them all.

Exercise 7.2

Find all the directed cycles contained in the regulatory network shown in Fig. 7.9 of this chapter. Also find the minimum FVS.

Exercise 7.3

The minimum FVS for the regulation network in Fig. 7.2 of this chapter is $\{A\}$. The dynamical system corresponding to this network is given by

$$\frac{du_A}{dt} = f_A(u_A, u_C) - d_A u_A$$
$$\frac{du_B}{dt} = f_B(u_A) - d_B u_B$$
$$\frac{du_C}{dt} = f_C(u_A, u_B) - d_C u_C$$

where f_A, f_B, and f_C are arbitrary bounded and differentiable nonlinear functions, and the decaying terms $d_A u_A$, $d_B u_B$, and $d_C u_C$ are assumed to be proportional to the concentration of each molecule. Let the two time-dependent solutions of this dynamical system be $\boldsymbol{u}(t)$ and $\tilde{\boldsymbol{u}}(t)$, and let the difference between them be $\boldsymbol{w}(t) := \tilde{\boldsymbol{u}}(t) - \boldsymbol{u}(t)$. Suppose that $w_A(t) = \tilde{u}_A(t) - u_A(t) \to 0 \quad (t \to +\infty)$. In this case, write the differential equations satisfied by $w_B(t)$ and $w_C(t)$. Also, show the behavior of $w_B(t)$ and $w_C(t)$ in the limit $(t \to +\infty)$ as sufficient time passes.

Exercise 7.4

The regulation system in Fig. 7.2 of Chap. 3 has a minimum FVS of size 1 that contains any one of the four molecules. If this system has multiple equilibrium points, show that they can always be distinguished by one variable, and that they can be distinguished by any one variable (Hint: Show that if the values of one variable are the same in two observations of steady states, then the values of the remaining variables are also the same).

References

Akutsu T, Kuhara S, Maruyama O, Miyano S (1998) A system for identifying genetic networks from gene expression patterns produced by gene disruptions and overexpressions. Genome Inform 9:151–160

Festa P, Pardalos PM, Resende MGC (2000) Feedback set problems. In: Du D-Z, Pardalos PM (eds) Handbook of combinatorial optimization, supplement vol. A. Kluwer Academic Publishers, pp 209–259

Fiedler B, Mochizuki A, Kurosawa G, Saito D (2013) Dynamics and control at feedback vertex sets. I: informative and determining nodes in regulatory networks. J Dyn Diff Equat 25:563–604

Foias C, Temam R (1984) Determination of the solutions of the Navier-Stokes equations by a set of nodal values. Math Comput 43:117–133

Imai KS, Levine M, Satoh N, Satou Y (2006) Regulatory blueprint for a chordate embryo. Science 312:1183–1187

Kalman RE (1963) Mathematical description of linear dynamical systems. J Soc Indust Appl Math Ser A 1:152–192

Kobayashi K, Maeda K, Tokuoka M, Mochizuki A, Satou Y (2018) Controlling cell fate specification system by key genes determined from network structure. iScience 4:281–293

Kobayashi K, Maeda K, Tokuoka M, Mochizuki A, Satou Y (2021) Controlling dynamics of a gene regulatory network of a chordate embryo using a structural theory. Sci Rep 11:4001

Lin CT (1974) Structural controllability. IEEE Trans Automat Control Ac19:201–208

Liu YY, Slotine JJ, Barabasi AL (2011) Controllability of complex networks. Nature 473:167–173

Mirsky HP, Liu AC, Welsh DK, Kay SA, Doyle FJ III (2009) A model of the cell-autonomous mammalian circadian clock. Proc Natl Acad Sci USA 106:11107–11112

Mochizuki A (2008) Structure of regulatory networks and diversity of gene expression patterns. J Theor Biol 250:307–321

Mochizuki A, Fiedler B, Kurosawa G, Saito D (2013) Dynamics and control at feedback vertex sets. II: a faithful monitor to determine the diversity of molecular activities in regulatory networks. J Theor Biol 335:130–146

Mochizuki A, Saito D (2010) Analyzing steady states of dynamics of bio-molecules from the structure of regulatory networks J Theor Biol 266:323–335

Zanudo JGT, Gang Y, Albert R (2017) Structure-based control of complex networks with nonlinear dynamics. Proc Natl Acad Sci USA 114(28):7234–7239

Chapter 8
Chemical Reaction Systems I: Control by Enzymes

Abstract In the previous chapter, we learned that an important aspect of the dynamical behavior of regulatory systems is determined by the structure of the regulatory network. In this and the next chapter, we will introduce structural theory for another class of networks in the life sciences, reaction networks. We will show that various aspects of the behavior of a reaction system are determined from the structure of the reaction network. In this chapter, we will focus on the response of the system when the parameters change, for example, the changes in the concentration of chemical substances when the amount or activity of the enzyme changes. The molecular species that show concentration changes to a change in enzyme activity are determined solely from the structure of the network.

Keywords Reaction network · Metabolic pathway · Enzyme regulation · Structural sensitivity analysis · Law of localization · Buffering structure

8.1 Chemical Reaction Network

The functions of cells as living organisms are fundamentally generated by the chemical reactions taking place in cells. A large number of chemical reactions occur within cells, and these are connected by sharing substrates and products, forming a network. For example, Fig. 8.1 shows the central metabolic system, which is a universal reaction pathway shared by various species from bacteria to humans. Through the reactions of the central metabolic system, sugar (glucose) is broken down into smaller substances, and chemical energy is extracted as ATP in the process. In Fig. 8.1, each arrow represents a state transition of chemical substances caused by a chemical reaction. The substance indicated at the tail of the arrow is the material, or substrate, of the reaction, and the substance indicated at the head of the arrow is the product of the reaction. The rate of a biochemical reaction is regulated by an organic catalyst called an enzyme that is specific to that reaction. It is known that the many chemical reactions occurring in living organisms are connected in a chain-like manner, sharing substrates and products, to form a huge network (see Fig. 1.2 in Chap. 1). It is believed that the physiological functions of the cell arise from the dynamics of this entire system. It is also believed that physiological

A. Mochizuki, *Structural Theories for Dynamical Networks in Biology*, Theoretical Biology, https://doi.org/10.1007/978-981-96-6026-1_8

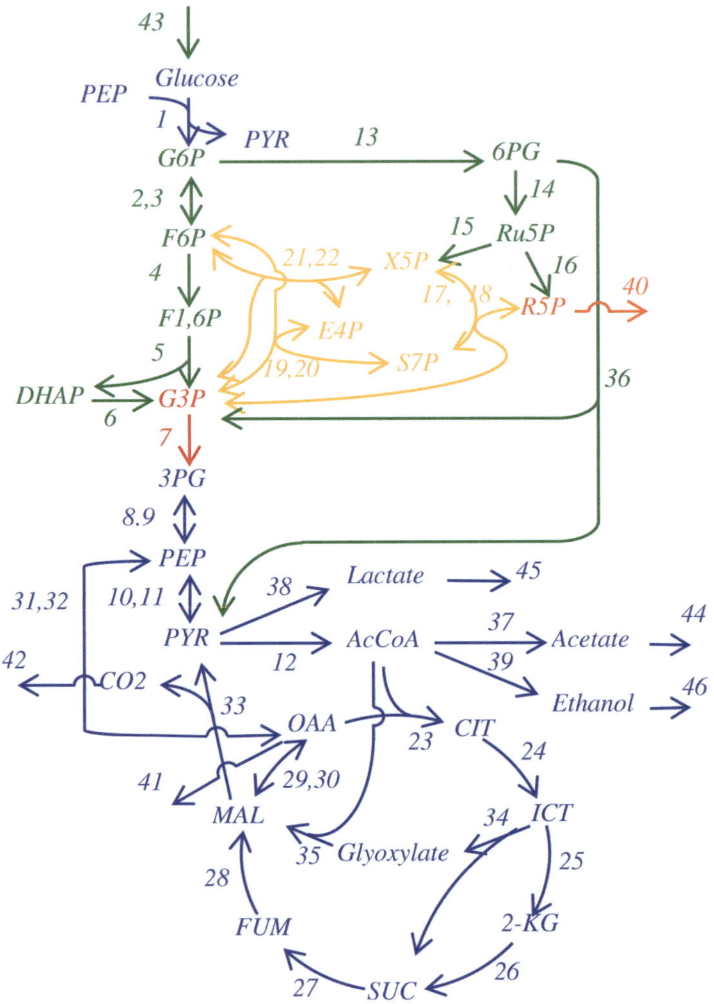

Fig. 8.1 Reaction network of the central metabolic system of bacteria (*E. coli*), including 28 chemical substances and 46 chemical reactions. Each arrow indicates a chemical reaction. The numbers accompanying the arrows distinguish the chemical reactions, and the letters at both ends of the arrows indicate the metabolites. The color corresponds to Fig. 8.13. (Based on Ishii et al. 2007, prepared by Okada and Mochizuki 2016)

functions are regulated by changes in the amount or activity of the enzymes that mediate the reactions.

The information on biochemical reactions that have been identified in the history of biology has gradually accumulated to form large-scale databases, and complex networks like the one shown in Fig. 8.1 can be seen in various situations. This is a network that has been naturally constructed by linking the identified reactions that share products or substrates. However, while it is a compilation of information on

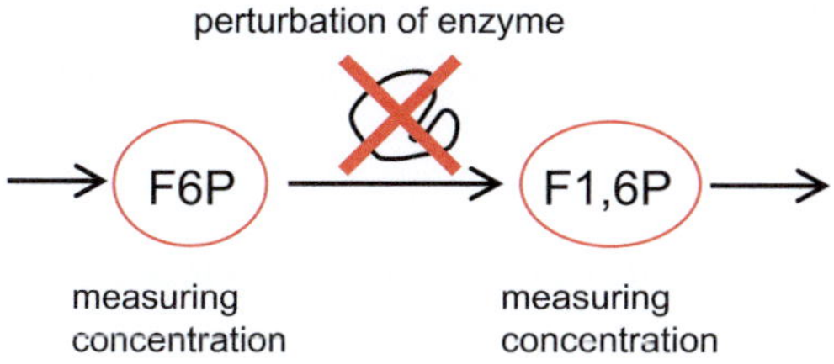

Fig. 8.2 Example of perturbation and response to a chemical reaction system. Perturbations are applied by knocking down or overexpressing enzymes of a reaction in the system, and the concentration of metabolites is measured

known reactions, surprisingly little is known about the dynamics of chemical substances based on this reaction network. In addition, it is not guaranteed that our current compilation of knowledge is complete.

One way to understand the behavior of a reaction system is to experimentally perturb the system and measure the response. Specifically, each of enzymes regulating reactions in a network is perturbed by knockdown or overexpression, and changes in the concentration of chemical substances are measured as a response to the perturbation. Since the 2000s, methods to measure many metabolites in a system at once using mass spectrometry have been actively used and are called metabolome analysis. These experiments are not merely measuring the response to perturbations. They can be interpreted as pseudo-control experiments by artificially inducing changes in enzyme activity that organisms actually perform to control physiological functions.

On the other hand, it has been said that the interpretation of the results obtained from such perturbation experiments is extremely difficult. For example, there is a pioneering study in which various enzymes were knocked down in the central metabolic system of *E. coli*, and changes in the concentration of metabolites were measured by metabolome analysis (Ishii et al. 2007). In this study, in many knockdown experiments, only a very small number of metabolites showed changes in concentration, and most of the metabolites showed no significant change. Even when changes were observed, the response pattern was difficult to understand intuitively. Let's take one specific example. Figure 8.2 shows part of the central metabolic system. As shown in the figure, one of the enzymes in a series of reactions was knocked down. In this case, the concentration of the product of the perturbed reaction almost did not change, but rather the concentration of the substrate increased. In response to these results, Ishii et al. considered that in actual biological metabolic systems, there may be unknown reactions that are not recorded in the database and that these bypass reactions may be creating the robustness of the response to enzyme perturbations.

But are these intuitive arguments correct? Rather, we need a guideline for discussing the results of perturbation experiments and the control of the system by enzymes based on solid logic, not intuition. In the following, we present a theoretical method for determining the response of a system to changes in the activity or quantity of an enzyme, based solely on the structure of the network.

8.2 Dynamics of Chemical Reaction Systems

The theory of the dynamics of chemical reaction systems has a long history; however, the basic idea is simple. That is, the change in concentration of each chemical substance in the system with time is determined by the difference between the influx and efflux caused by reactions.

For example, in the reaction system shown in Fig. 8.3, the concentration of chemical substance A increases with the reaction rate of reaction 1 and decreases with the reaction rate of reaction 2. For each chemical substance, this relationship between the time change in concentration and the difference between the reaction rates can be expressed by a system of differential equations:

$$\frac{du_A}{dt} = r_1 - r_2 \tag{8.1a}$$

$$\frac{du_B}{dt} = r_2 - r_3 \tag{8.1b}$$

where u_m ($m = A, B$) is the concentration of chemicals in the system and r_j ($j = 1, 2, 3$) is called the reaction rate, indicating the rate per unit time for each reaction. (8.1) may be written in the following form:

$$\frac{d}{dt}\begin{pmatrix} u_A \\ u_B \end{pmatrix} = \begin{pmatrix} 1 & -1 & 0 \\ 0 & 1 & -1 \end{pmatrix} \begin{pmatrix} r_1 \\ r_2 \\ r_3 \end{pmatrix} \tag{8.2}$$

In general, the dynamics of a chemical reaction system can be described in the following form:

$$\frac{d}{dt}\begin{pmatrix} u_1 \\ \vdots \\ u_M \end{pmatrix} = \begin{pmatrix} \nu_{11} & \cdots & \nu_{1J} \\ \vdots & & \vdots \\ \nu_{M1} & \cdots & \nu_{MJ} \end{pmatrix} \begin{pmatrix} r_1 \\ \vdots \\ r_J \end{pmatrix} \tag{8.3}$$

where $(u_1, \cdots, u_M)^T$ is the vector of concentration of chemical substances in the system and $(r_1, \cdots, r_J)^T$ is the reaction rate vector. $\nu = (\nu_{mj})$ is called the

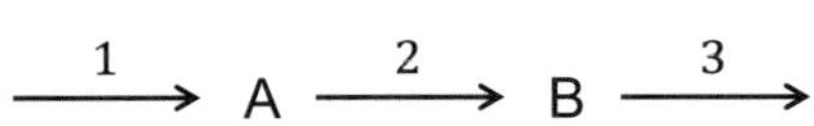

Fig. 8.3 Example of a chemical reaction system, corresponding to Fig. 8.2. Corresponding to Fig. 8.2, where A and B are the two chemical substances in the system and 1, 2, and 3 are the three reactions

"stoichiometric matrix," where each row corresponds to each chemical and each column to each reaction. Each component is usually written as an integer, indicating the number of molecules of each chemical that will decrease or increase for each reaction to occur once. ν is determined from the reaction network. The reaction rate function r_j is usually modeled using a mass action ($r_2 = k_2 u_A$) or a Hill function ($r_2 = k_2 u_A^n / (K^n + u_A^n)$).

Although not discussed in detail in this chapter, there are theories that facilitate mathematical understanding by assuming a mass-action kinetics for the reaction rate function (Clarke 1980). For example, Feinberg proved that if the graph of a chemical reaction network is strongly connected and the graph index, called "Deficiency," is zero, then it is a sufficient condition for the unique existence of a stable stationary solution (Feinberg 2019).

8.3 Structural Theory for Chemical Reaction Systems—Structural Sensitivity Analysis

Let us focus on a steady state of the reaction system in Eq. (8.3). Considering the steady state is a natural assumption, based on the fact that the dynamics of chemical reactions are much faster than the dynamics of gene expression, which governs the amount of enzymes, etc. We will now introduce a method to determine the steady-state response of a reaction system to perturbations of reaction enzymes from network information alone (Mochizuki and Fiedler 2015; Okada and Mochizuki 2016, 2017). The analysis to determine the response to a perturbation is generally called sensitivity analysis. The method presented here is called "structural sensitivity analysis" in the sense that it performs sensitivity analysis based only on the network structure. Note that, although we use the expression "perturbation of the enzyme," the sensitivity to any parameter that affects the reaction rate is given by the same formula.

In the following discussion, we do not assume any specific form of function for the reaction rate r_j. Only two obvious facts are used. (1) Concentration dependence: The rate of each reaction depends on the concentration of the substrate chemical and the regulatory chemical. (2) Reaction parameters: Each reaction rate has a single parameter corresponding to the activity or amount of the enzyme. For (1), it is known that chemical substances other than the substrate can regulate enzyme activity, which is called allosteric regulation. The concentration of the substrate chemicals is naturally included in the argument of the reaction rate function, but in addition, the regulator of known allosteric regulation is also an argument of the reaction rate function. In the following, when allosteric regulation is considered, it is explicitly indicated on the network (e.g., Chap. 9, Figs. 9.5 and 9.6). These are all represented by the non-zero partial derivative $\partial r_j / \partial u_m$ of the reaction rate r_j with respect to each chemical concentration u_m.

In the main text, I will only explain the methods and results of the structural sensitivity analysis. For the derivation of the method, see Mathematical Method 8A. The following two-step calculation allows us to determine the response of all chemical concentrations and all reaction fluxes to perturbations in any reaction parameter in the system at once. Here, we consider the case where there are no conserved quantities in the system. This corresponds to the case where the stoichiometric matrix is full rank (rank($\boldsymbol{\nu}$) $= M$) (8A mathematical method). In a first step, we construct the matrix

$$\mathbf{A} := \left(\begin{array}{ccc|ccc} \frac{\partial r_1}{\partial u_1} & \cdots & \frac{\partial r_1}{\partial u_M} & & & \\ \vdots & & \vdots & -\boldsymbol{c}^1 & \cdots & -\boldsymbol{c}^N \\ \frac{\partial r_J}{\partial u_1} & \cdots & \frac{\partial r_J}{\partial u_M} & & & \end{array} \right) \tag{8.4}$$

from the chemical reaction network. The left part $\partial r_j/\partial u_m$ of matrix $\mathbf{A}$ is the partial derivative of the reaction rate with respect to the substrate concentration, and the right part $\boldsymbol{c}^n$ represents the basis vector of the kernel space $\ker(\boldsymbol{\nu}) = \{\boldsymbol{c} \in \mathbb{R}^J | \boldsymbol{\nu}\boldsymbol{c} = \mathbf{0}\}$ of stoichiometric matrix $\boldsymbol{\nu}$. Each row of matrix $\mathbf{A}$ corresponds to each reaction ($j = 1, \cdots, J$), and the first half of the columns correspond to each chemical species ($m = 1, \cdots, M$). In the second half of the columns of the matrix $\mathbf{A}$, $-\boldsymbol{c}^n$ is arranged in each column ($n = 1, \cdots, N$), where N is the dimension of the kernel space of $\boldsymbol{\nu}$. (When there are no conserved quantities, $J = M + N$.) The left part $\partial r_j/\partial u_m$ shows the dependence of each reaction on each substrate and is zero when reaction j is independent of substrate m. When reaction j depends on substrate m (e.g., when it is the substrate of a reaction), it takes a non-zero value, which is explicitly indicated below by writing $r_{jm}(=\partial r_j/\partial u_m)$. The right part $\boldsymbol{c}^n$, that is, the basis vector of kernel space of the matrix $\boldsymbol{\nu}$, corresponds to the flow path at steady state determined from the network.

For example, for the reaction system shown in Fig. 8.3 (Eq. (8.2)), note that the stoichiometric matrix $\boldsymbol{\nu}$ is full rank (rank($\boldsymbol{\nu}$) = 2) and dim(ker($\boldsymbol{\nu}$)) = 3 − rank ($\boldsymbol{\nu}$) = 1, the matrix $\mathbf{A}$ is given by

$$\mathbf{A} = \begin{pmatrix} 0 & 0 & -1 \\ r_{2A} & 0 & -1 \\ 0 & r_{3B} & -1 \end{pmatrix} \tag{8.5}$$

Each row corresponds to reaction 1 to reaction 3, and each column corresponds to chemicals A and B and the basis vector of ker($\boldsymbol{\nu}$). The non-zero component r_{2A} in row 2, column 1 indicates that reaction 2 depends on the concentration of chemical A (substrate), and the zero component in row 2, column 2 indicates that reaction 2 does not depend on the concentration of chemical B (product). The non-zero component r_{3B} in row 3, column 2 indicates that reaction 3 depends only on the concentration of chemical B. Reaction 1 does not depend on the concentration of either chemicals. As will become clear below, in this analysis, the pattern of the 0 components as well as the non-zero components is important information. The basis vector of ker($\boldsymbol{\nu}$) in column 3 corresponds to the path through reactions 1, 2, and 3, from the inflow to the outflow of the system. In this way, the patterns of zero and non-zero components of matrix A directly reflect the topology of the network.

In the second step, we compute the inverse of matrix **A**. The change in the concentration of each chemical (or the change in the rate of each reaction) upon perturbation of each reaction parameter is given below as each component of the inverse matrix:

$$\mathbf{S} \equiv \begin{pmatrix} \delta_1 u_1 & \cdots & \delta_J u_1 \\ \vdots & & \vdots \\ \delta_1 u_M & \cdots & \delta_J u_M \\ \hline \delta_1 \mu^1 & \cdots & \delta_J \mu^1 \\ \vdots & & \vdots \\ \delta_1 \mu^N & \cdots & \delta_J \mu^N \end{pmatrix} \propto -\mathbf{A}^{-1} \tag{8.6}$$

where $\delta_j u_m$ denotes the change in concentration of chemical m when the parameter of reaction j is perturbed. $\delta_j \mu^n$ denotes the change in the coefficient of the steady flow along $\boldsymbol{c}^n$ when the parameter of reaction j is perturbed. More directly, the change in the reaction rate vector $\boldsymbol{r}$ to the perturbation of the reaction parameter j is given by:

$$\delta_j \boldsymbol{r} = \sum_{n=1}^{N} \delta_j \mu^n \boldsymbol{c}^n \tag{8.7}$$

Here, each component of the sensitivity matrix $\boldsymbol{S}$ is either zero or nonzero, which can be determined from the pattern of the zero (or nonzero) components of the matrix **A**. As stated above, the pattern of the zero (or nonzero) components of matrix **A** is determined solely from the structure of the reaction network. Therefore, from the network structure alone, it is possible to determine the response of the system to perturbations in the reaction constants, i.e., the presence or absence of changes in the concentration of each chemical and in the rate of each reaction. Furthermore, by assuming that each reaction rate is a monotonically increasing or decreasing function of the concentration of each chemical (a natural assumption), in many cases the sign of the change, that is, the increase or decrease in concentration or rate, can also be determined from the network structure alone.

For the dynamics shown in (8.2) (matrix **A** in (8.5)), the sensitivity matrix **S** is given by (Exercise 8.3):

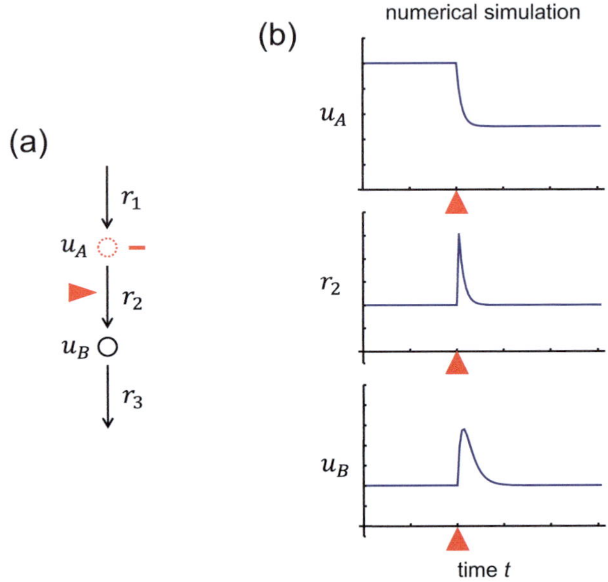

Fig. 8.4 An example of response and a perturbation to a chemical reaction system. (**a**) Red arrowhead indicate the reaction in which the parameter is increased, and the red dotted circle indicates the chemical showing decreases in concentration. Black circles and black arrows indicate no change in the steady state before and after the perturbation. (**b**) Temporal changes in the chemical concentration and reaction rate. The red arrowheads indicate the time at which the perturbation was applied

$$\mathbf{S}=\begin{pmatrix} 1/r_{2A} & -1/r_{2A} & 0 \\ 1/r_{3B} & 0 & -1/r_{3B} \\ 1 & 0 & 0 \end{pmatrix}. \tag{8.8}$$

Here each column represents the results of perturbing the parameters of each of the three reactions 1–3, and each row represents the response of each chemical and the steady flow, i.e., row 1 shows the response of chemical A, row 2 shows the response of chemical B, and row 3 shows the response of the steady flow.

Figure 8.4 illustrates the results of increasing the parameter k_2 associated to reaction 2. This perturbation refers to an experiment in which the activity or expression level of the enzyme in reaction 2 is increased. One might expect an increase in the concentration of the product chemical B, or an increase in the rate of the perturbed reaction as the response to the perturbation. However, the conclusion of the structural sensitivity analysis is different. As shown in column 2 of (8.8), the only response to the increase of the enzyme activity in reaction 2 is decrease of substrate A. Neither the concentration of substrate B nor any of the three reaction rates change at steady state in response to the perturbation.

Let us confirm these results by performing numerical simulations. The theory does not depend on the mathematical form of the function, but when we choose an arbitrary increasing function of substrate for each reaction rate and calculate the dynamics, we should observe the behavior shown in Fig. 8.4b. Soon after increasing the enzyme activity, the reaction rate r_2 and the concentration of the product chemical u_B increase, but this is only temporary. Immediately the reaction rate r_2 decreases and converges to the level before the perturbation. At the same time, the

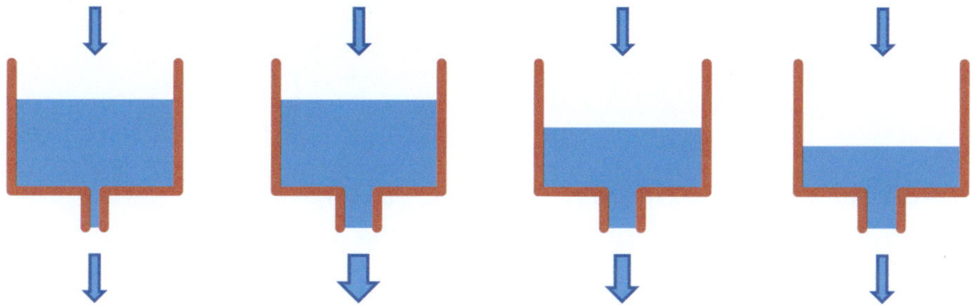

Fig. 8.5 Intuitive explanation for the fact that only the substrate concentration changes as a result of the enzymatic perturbation in Fig. 8.4

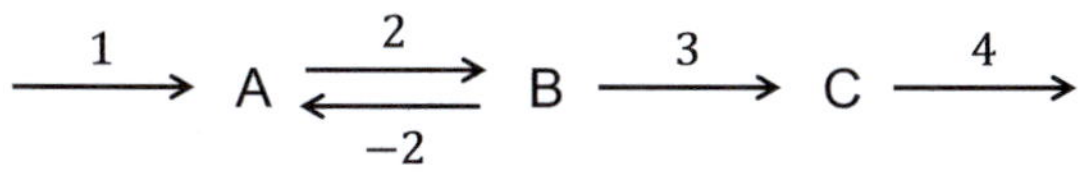

Fig. 8.6 Example of a chemical reaction system, where *A* to *C* represent the three chemicals in the system, 1, 2, −2, 3, and 4 represent the five reactions, and the arrows indicate the state transition due to these reactions

concentration of the substrate chemical, u_A, decreases. Focusing on the steady state, only the concentration of the substrate chemical *A* changes before and after the perturbation. Qualitatively the same behavior is observed regardless of the form of reaction rate function r_j (as long as it monotonically increases with substrate concentration). We may interpret that the change in the reaction parameter is compensated by the change in the concentration of the substrate chemical, and the concentrations and reaction rates downstream do not change at all eventually (Fig. 8.5).

Intuitively, it is easy to understand by considering a water tank with an inflow and an outflow (Fig. 8.5). Increasing enzyme activity corresponds to opening the faucet and enlarging the outlet. As the outlet is enlarged, the amount of water flowing out naturally increases. The increase in outflow will lead to a decrease in the water level, which will lead to a decrease in the water pressure at the bottom of the tank. The decrease in water pressure causes the outflow to decrease. This process will stop the decrease in the water level eventually when the inflow and outflow are in balance. This is the new steady state, where the only change observed before and after the operation (enlargement of the outlet) is a decrease in the water level. In other words, the behavior observed in the mathematical analysis described above is a very natural phenomenon. The important point is that, without having to assume a function and perform numerical calculations, or even thinking about an intuitive explanation, it is possible to determine the qualitative change in the steady state using simple matrix calculations.

Let us give another example: For the chemical reaction system shown in Fig. 8.6, the dynamics of the concentrations of *A*, *B*, and *C* are given by the differential equation system:

$$\frac{d}{dt}\begin{pmatrix} u_A \\ u_B \\ u_C \end{pmatrix} = \begin{pmatrix} 1 & -1 & 1 & 0 & 0 \\ 0 & 1 & -1 & -1 & 0 \\ 0 & 0 & 0 & 1 & -1 \end{pmatrix}\begin{pmatrix} r_1 \\ r_2 \\ r_{-2} \\ r_3 \\ r_4 \end{pmatrix} \tag{8.9}$$

Note that stoichiometric matrix $\boldsymbol{\nu}$ is full rank, rank$(\boldsymbol{\nu}) = 3$ and dim $(\ker(\boldsymbol{\nu})) = 5 -$ rank $(\boldsymbol{\nu}) = 2$, so ker$(\boldsymbol{\nu})$ has two basis vectors. Matrix $\mathbf{A}$ is given by:

$$\mathbf{A} = \begin{pmatrix} 0 & 0 & 0 & -1 & 0 \\ r_{2A} & 0 & 0 & -1 & -1 \\ 0 & r_{-2B} & 0 & 0 & -1 \\ 0 & r_{3B} & 0 & -1 & 0 \\ 0 & 0 & r_{4C} & -1 & 0 \end{pmatrix} \tag{8.10}$$

Here, each row corresponds to each reaction (1, 2, − 2, 3, 4), and each column corresponds to chemical A, B, C or to basis vectors of ker$(\boldsymbol{\nu})$. r_{jm}'s are nonzero components of the partial derivative of the reaction rate r_j with respect to the chemical concentrations u_m. For example, r_{2A} in (2, 1) indicates that the rate of reaction 2 depends on the concentration of A. The two basis vectors correspond to the pathways that run through reactions 1, 2, 3, 4 from inflow to outflow (column 4) and to the pathways of the cycle in the system consisting of reactions 2 and −2 (column 5).

Sensitivity matrix $\mathbf{S}$ in this case is given by (Exercise 8.4):

$$\mathbf{S} = \begin{pmatrix} \frac{r_{-2B} + r_{3B}}{r_{2A}r_{3B}} & -\frac{1}{r_{2A}} & \frac{1}{r_{2A}} & -\frac{r_{-2B}}{r_{2A}r_{3B}} & 0 \\ \frac{1}{r_{3B}} & 0 & 0 & -\frac{1}{r_{3B}} & 0 \\ \frac{1}{r_{4C}} & 0 & 0 & 0 & -\frac{1}{r_{4C}} \\ 1 & 0 & 0 & 0 & 0 \\ \frac{r_{-2B}}{r_{3B}} & 0 & 1 & -\frac{r_{-2B}}{r_{3B}} & 0 \end{pmatrix} \tag{8.11}$$

The responses of the system to the increase in the parameter of each reaction are obtained using (11) as follows (Fig. 8.7):

Reaction 1: The concentrations of A, B, and C increase. The fluxes of the pathways through the system and the cycle pathway consisting of reactions 2 and −2 increase.

Reaction 2: The concentration of A decreases. The concentrations of the other chemicals are unchanged. Fluxes of all reactions are unchanged.

Reaction −2: The concentration of A increases, whereas those of the other chemicals remain unchanged. The fluxes of the cycle pathway consisting of reactions 2 and−2 increase.

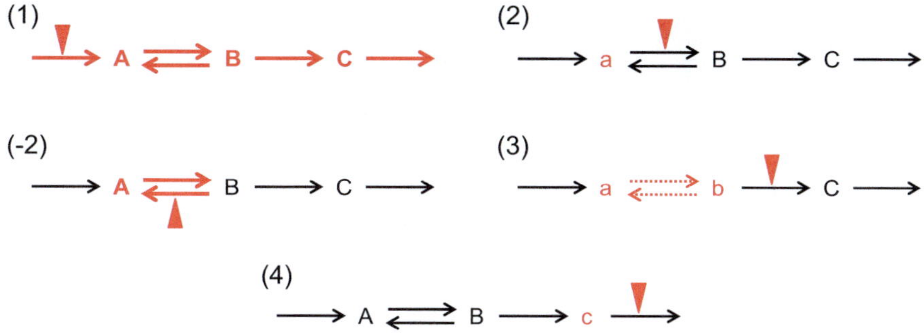

Fig. 8.7 Response patterns of the chemical reaction system (Fig. 8.6) to increasing parameter for each reaction. The results of the structural sensitivity analysis (8.11) are shown in the figure. Red arrowheads indicate perturbed reactions. The red letters indicate changes in the concentration of the chemical—red uppercase letters indicate increase in concentration and red lowercase letters indicate decrease in concentration. Red arrows indicate changes in the reaction rate—red thick arrows indicate increase in reaction rate and red dotted arrows indicate decrease in reaction rate. Black letters and black arrows indicate chemicals and reactions that do not change before and after the perturbation

Reaction 3: The concentrations of A and B decrease, whereas that of C does not change. The flux of the cycle pathway consisting of reactions 2 and−2 decreases.
Reaction 4: The concentration of C decreases, whereas those of the other chemicals remain unchanged. Fluxes of all reactions are unchanged.

Note that perturbation responses of reactions 2 and −2 are asymmetrical and that the effect of perturbation of reaction 3 is transmitted two steps upstream. In this way, perturbations of the reaction parameters generally show characteristic response patterns, depending on the topology of the network and the position of the perturbed reaction.

8.4 Application (1): Plant Metabolic Systems

Below, we will introduce an example of applying structural sensitivity analysis to an actual biological system. Ferjani et al. found an abnormality in the metabolic system in the mutant they were analyzing (Ferjani et al. 2018). Specifically, the sucrose production is significantly reduced in the early development of the mutant plants. Sucrose is a sugar that plays an important role in intercellular transport in plants. On the other hand, it was found that the function of the enzyme H^+-pyrophosphatase, which degrades pyrophosphate (PPi), is lost in this mutant. PPi is naturally formed by the hydrolysis of ATP into AMP in many biological processes in cells, and the accumulation of them to toxic levels disrupts several common biosynthetic pathways and is lethal. Figure 8.8 shows the pathway by which sucrose is synthesized from fructose in early plant development. Pyrophosphate is synthesized as a by-product of

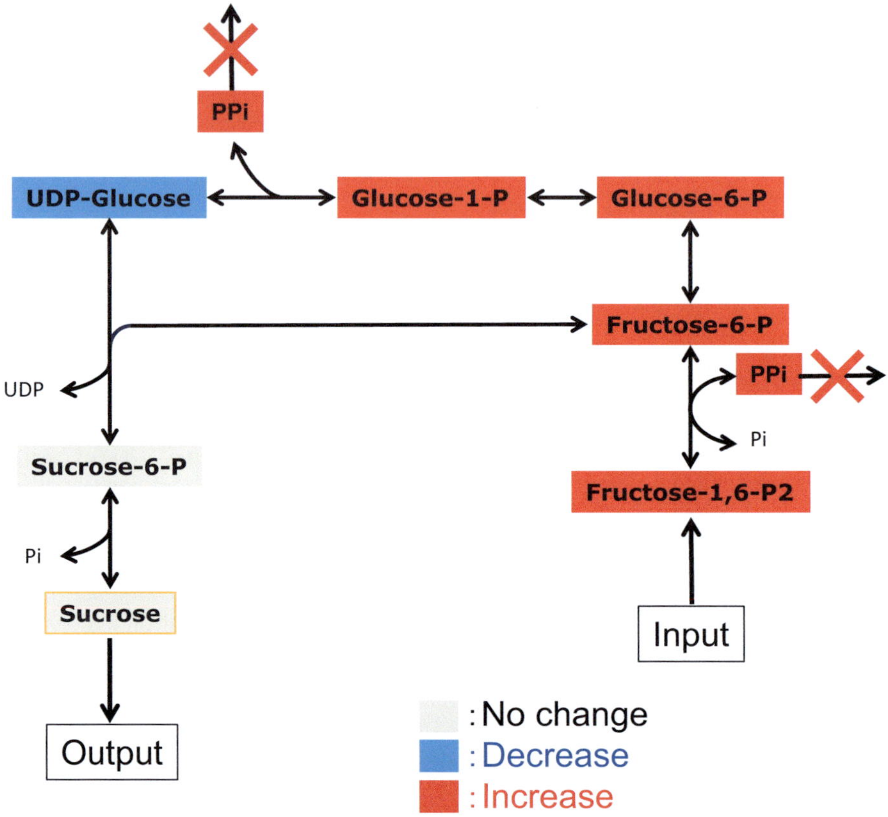

Fig. 8.8 Metabolic pathways involved in the synthesis of sucrose from fructose during early plant development. Changes in the concentration of each metabolite upon knockdown of the PPi-degrading enzyme were determined by structural sensitivity analysis and are indicated by the color of each chemical (Ferjani et al. 2018). Red: increase, blue: decrease, gray: no change

two reversible reactions on this pathway. Based on these findings, the following story was initially assumed. In individuals in which the pyrophosphate degrading enzyme is disrupted, the concentration of pyrophosphate would increase, which would make slowing down of the forward direction of the two reversible reactions on the sucrose synthesis pathway. Therefore, the rate of sucrose synthesis downstream of the pathway may decrease. At first glance, this story sounds reasonable.

To verify this story, Ferjani et al. applied structural sensitivity analysis to the sucrose synthesis pathway in plants. Figure 8.8 shows the changes in the concentration of each chemical as a result of disrupting PPi-degrading enzymes (pyrophosphatase), indicated by the colors. Contrary to expectations, disruption of pyrophosphatase does not result in a decrease in sucrose. The concentration of pyrophosphate indeed increases by the destruction of pyrophosphatase. The concentrations of the four sugars upstream in the pathway also increase, and the concentration of UDP-glucose, which is the product of the reaction releasing pyrophosphate as a by-product, decreases. So far, the intuitive story is true. However, the concentration of sucrose-6P, product of complex formation of fructose-6P and UDP-glucose, does not change. The concentration of sucrose, which is the downstream of sucrose-6P, is also unchanged. As long as we consider based on the reaction network in Fig. 8.8, this result is always the case, regardless of the choice of reaction rate functions or parameters. On the other hand, in actual plants, a decrease in sucrose concentration has certainly been observed in the mutant of pyrophosphatase. This means that we must revise the network structure in Fig. 8.8 and reconsider the problem.

The double arrow in Fig. 8.9 is an example of a modification of the network. A network with one additional reaction was assumed, and the responses to the destruction of pyrophosphatase were determined by structural sensitivity analysis using the modified network. In this case, the response of the concentration of each sugar is

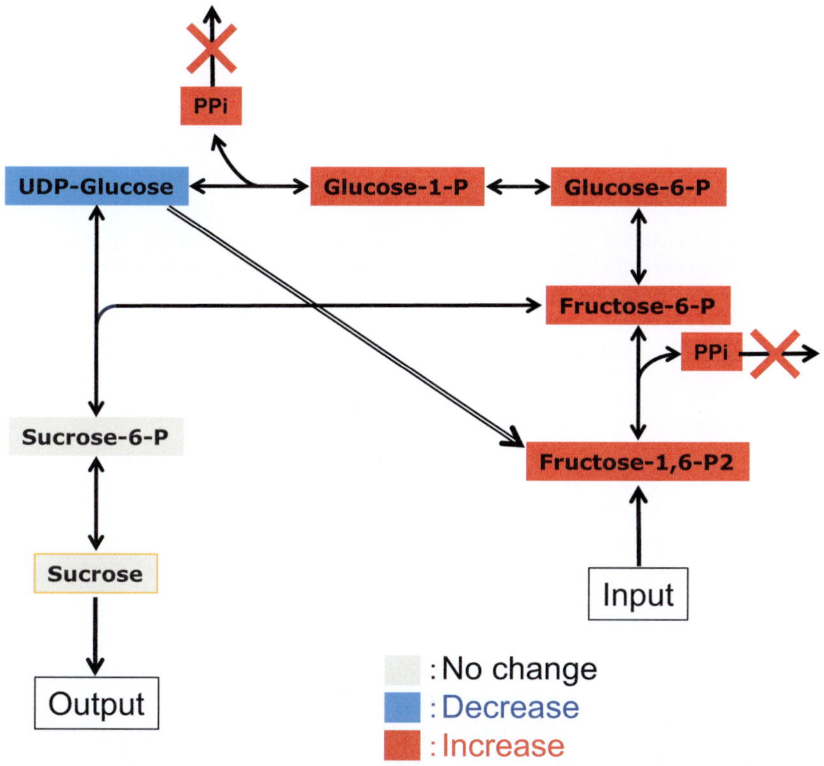

Fig. 8.9 Response of the modified network with one additional reaction to the perturbation of PPi degrading enzymes. Double arrows indicate a newly added reaction. The addition of this reaction does not change the qualitative response of the system to the PPi-degrading enzyme perturbation at all

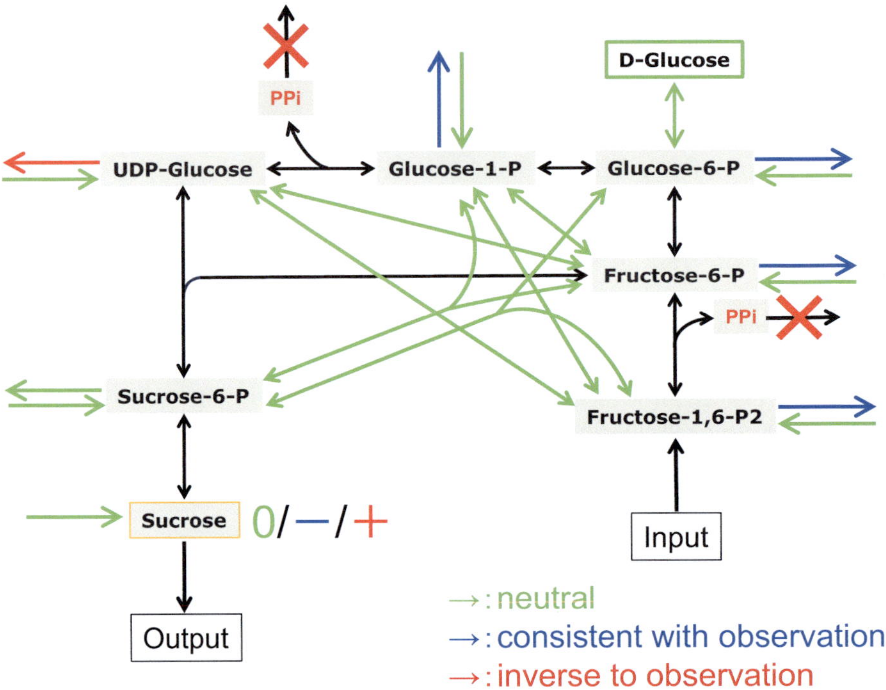

Fig. 8.10 Effect of adding various reactions to the plant metabolic pathway (Fig. 8.8) on the sucrose response. Green: The addition of these reactions does not change the response of sucrose concentration upon the destruction of PPi-degrading enzymes. Arrows with arrowheads at both ends indicate that either directional or reversible reactions do not change the results. Blue: When one of these four reactions is added, the concentration of sucrose in the network decreases in response to the destruction of PPi-degrading enzymes. Red: When this reaction is added, the concentration of sucrose in the network rather increases in response to the destruction of the PPi-degrading enzyme

exactly the same as before the modification was applied. That is, this modification cannot explain the phenomenon. On the other hand, there are various other possible modifications. Ferjani et al. added each of these candidate reactions to the network and analyzed the response to pyrophosphatase disruption in these modified networks. Figure 8.10 summarizes these results.

Most of the reactions added do not change the system's response to the destruction of pyrophosphatase. Only the modified network with the addition of one of the four reactions indicated in blue causes the observed behavior, namely a decrease in sucrose concentration in response to the destruction of pyrophosphatase. On the other hand, the addition of the reaction indicated in red causes the opposite of the observed behavior, namely an increase in sucrose concentration in response to the destruction of pyrophosphatase.

From these results, Ferjani et al. predict that in the actual plant metabolism system, at least one of the reactions indicated in blue exists and the reaction shown in red does not exist. This prediction was confirmed to be true. It was verified by experimental measurement that the concentration of the products of the reactions indicated in blue changed in the mutant. Thus, by combining the structural sensitivity analysis with the knowledge of the network and the observed behavior of the system, it is possible to predict the reaction that should be present and specifically locate it in the system.

8.5 Law of Localization in Reaction Systems

In the previous sections, we have shown that the qualitative response of a chemical reaction system to a perturbation of an enzyme can be determined solely from the structure of the network. In each case, the effect of the perturbation of the reaction enzyme was confined to some chemical concentration and some reaction rates at steady state, i.e., within a finite range on the network. It can be immediately stated that the common intuitive argument that "upstream perturbations affect downstream" in a reaction network is wrong.

Such behavior is universally observed. That is, the effect of a perturbation on a reaction parameter is often confined to a limited part on the reaction network. There is a clear boundary between the part of the network that is affected by the perturbation and the part that is unaffected. Okada and Mochizuki discovered the mathematical laws that govern the behavior of reactive systems behind this phenomenon (Okada and Mochizuki 2016, 2017).

Consider an arbitrary substructure in a reaction network. This means selecting a subset of chemicals and a subset of reactions. Here we impose the following condition: (i) *Reactions corresponding to arrows originating from chemicals in the chemical subset must be contained in the reaction subset*. This condition is called "output complete." After satisfying output complete, it is also possible to add other reactions to the subset. A substructure that satisfies (i) output complete and the following condition is called a "buffering structure": (ii) (*number of chemicals*) − (*number of reactions*) + (*number of cycles*) − (*number of conserved quantities*) = 0.

Theorem (law of localization)

> Any perturbations given into reaction parameters inside a buffering structure affect the chemical concentrations and reaction rates within the structure only at steady state. The steady state of the chemical concentrations and reaction rates outside the buffering structure is not affected at all by the perturbations inside the buffering structure.

The proof is given only by the pattern of the zero components of the matrix **A**. See 8B Mathematical Methods for details. Furthermore, the inverse of this theorem is

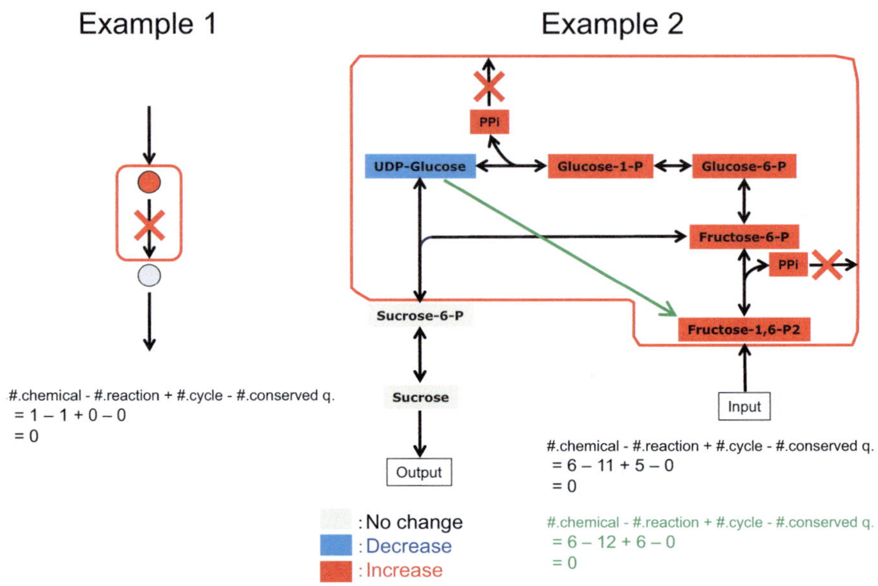

Fig. 8.11 Example of buffering structures. Corresponds to Figs. 8.2 (a) and 8.8 (b). Note that in (**b**), the PPi depicted in two places are the same chemical. The value of the equation remains 0 regardless of the presence or absence of the reaction indicated in green, so the buffering structure remains as it is

also true. That is, if the effects of a parameter perturbation are localized to a finite range on the network, then there will always be a corresponding buffering structure (Yamauchi et al. 2024).

The example in Fig. 8.11a is the simplest buffering structure. The structure containing one substrate and one reaction emanating from the substrate satisfies the condition of a buffering structure ($1 - 1 + 0 - 0 = 0$). In general, if there is only one reaction arrow emanating from a chemical, any perturbation applied to the reaction will affect only the substrate chemical. Figure 8.11b shows the case of the plant metabolism system: PPi and its degradation reaction are depicted in two places, but since they are the same chemical, the number of chemicals inside the red frame is 6. Before the modification (without including the green reaction), the number of reactions is 11 as the reversible reaction counts one forward and one backward. There are five reversible reactions in the red frame conforming small cycles. The substructure in the red frame satisfies the condition of a buffering structure ($6 - 11 + 5 - 0 = 0$). In other words, although we discussed the effect of perturbation to PPi-degrading enzymes in Sect. 8.5, not only that, but also the perturbation to any of the reaction parameters in the red frame in Fig. 8.11b has no effect on the concentration of sucrose-6P or sucrose.

In Fig. 8.10, various network modifications were made, but most of them did not change the sucrose response at all. This can be understood as follows. In Fig. 8.11, adding the reaction shown in green increases the number of reactions by one. However, at the same time, a new cycle structure is also created through the addition of the green reaction. Therefore, the value of the conditional expression remains 0 (=6 − 12 + 6 − 0), and the buffering structure remains unchanged.

There were only five network modifications resulting in new systems where the perturbation of PPi-degrading enzyme affected the concentration of sucrose (Fig. 8.10). All of these were additions of reactions emanating outward from the chemicals inside the buffering structure shown in Fig. 8.11b. According to the conditions for a buffering structure, we understand that only the addition of this kind of reaction is able to break the condition. In other words, the five reactions shown in blue or red in Fig. 8.11b are all of possible modifications that change the sucrose response, i.e., they are not only sufficient but also necessary conditions for altering sucrose response.

Note the fact that any substructure satisfying the above topological conditions by chance obtains the property of localized response to perturbations. Properties of robustness to enzyme fluctuations may arise simply from the topology of subnetworks, which could be realized just by chance in the evolution. Okada and Mochizuki created a network randomly to see how often buffering structures would be formed (Okada and Mochizuki 2017) and found that buffering structures appear frequently in random networks.

Let us illustrate the characteristic patterns exhibited by enzymatic perturbations to a chemical reaction system using another network. Multiple buffering structures may exist in a network, which can lead to more interesting patterns in the perturbation response. Figure 8.12 shows a hypothetical reaction system and the responses of chemical concentration and reaction rate to reaction perturbations within this system.

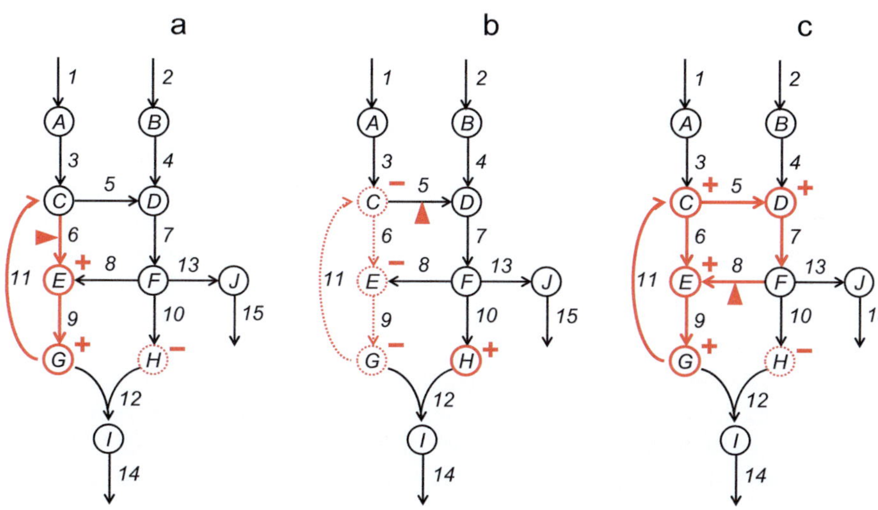

Fig. 8.12 Diversity of responses to enzymatic perturbations in a hypothetical chemical reaction system

These behaviors can be confirmed by numerical calculations, but none of them are obvious. For example, as shown in Fig. 8.12a, when the enzyme activity of reaction 6 is increased, the concentrations of E and G increase and the concentration of H decreases at steady state. On the other hand, the concentrations of the other chemicals do not change. Only the reaction rates of 6, 9, and 11 increase. This is because the substructure $\Gamma_1 = (\{E, G, H\}, \{6, 9, 11, 12\})$ satisfies the condition of a buffering structure ($3 - 4 + 1 - 0 = 0$). Next, in Fig. 8.12b, when the enzyme activity of reaction 5 is increased, a decrease in the steady-state concentrations of C, E, and G, an increase in the concentration of H, and a decrease in the reaction rates of 6, 9, and 11 are observed. This can be understood by the fact that the substructure $\Gamma_2 = (\{C, E, G, H\}, \{5, 6, 9, 11, 12\})$ is a buffering structure ($4 - 5 + 1 - 0 = 0$). Furthermore, as Fig. 8.12c shows, when the enzyme activity in reaction 8 is increased, an increase in C, D, E, and G, a decrease in H, and a decrease in the reaction rates of 5, 6, 7, 8, 9, and 11 at steady state are observed. This is due to the fact that the substructure $\Gamma_3 = (\{C, D, E, G, H\}, \{5, 6, 7, 8, 9, 11, 12\})$ is a buffering structure ($5 - 7 + 2 - 0 = 0$).

One may note that the three buffering structures listed here are in an inclusion relation with each other. That is, $\Gamma_1 \subset \Gamma_2 \subset \Gamma_3$. It has been proved that if there is an intersection between two buffering structures, the intersection is also a buffering structure. Therefore, when there are multiple buffering structures, they can have an inclusion relationship, or nested structure, with each other. As a result, when perturbations are applied to various locations in a chemical reaction system, the overall response shows a hierarchical nested pattern. The next section will show a remarkable example.

8.6 Application (2): Central Metabolic Pathway

Okada and Mochizuki analyzed responses of central metabolic pathway of *E. coli* (Fig. 8.1) to perturbations of reaction parameters systematically (Okada and Mochizuki 2016). Nonzero responses of the concentrations of chemicals to the perturbations of each reaction parameter in the system are summarized in Fig. 8.13, that shows the localized and hierarchical property in the patterns. When, for example, any one of the six reactions in the yellow box is perturbed, only three chemicals in the yellow box show changes in their concentration. Similarly, when reaction 7 (shown in red) is perturbed, only four chemicals in the red and yellow boxes show concentration change; whereas when reaction 40 (shown in orange) is perturbed, only four chemicals in the orange and yellow boxes show concentration change. When the reaction shown in blue is perturbed, there are 11 cases where only one chemical responds, and in other cases all blue chemicals respond. When the reaction shown in green is perturbed, green, yellow, red, orange, and blue chemicals

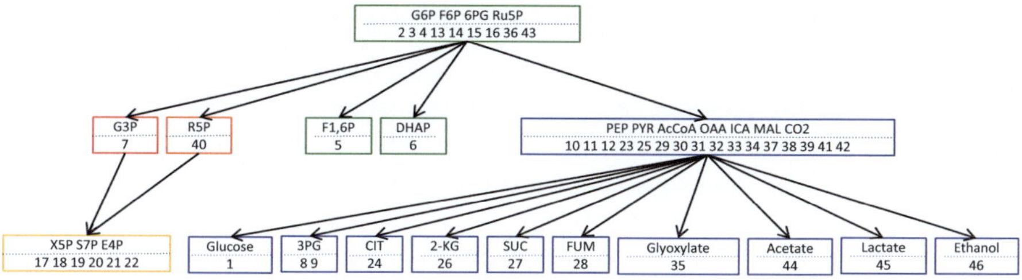

Fig. 8.13 The perturbation response network of the *E. coli* central metabolic pathway obtained from the structural sensitivity analysis. The numbers at the bottom of each box indicate the reaction parameters that were perturbed. The chemicals shown in each box and in all boxes followed by an arrow emanating downward from it indicate the chemicals whose concentration changes in response to a given perturbation. Colors correspond to those in Fig. 8.1. (Okada and Mochizuki 2016)

respond. Thus, not only the responses to perturbations of various reactions are respectively localized, the response ranges show inclusion relations with each other.

The hierarchy in responses patterns observed in the central metabolic pathway of *E. coli* can be understood as the nest of buffering structures. There are 17 buffering structures in this network, and they are nested in each other. In Fig. 8.13, a buffering structure corresponds to chemicals and reactions shown in each box and in the boxes followed by the downward arrows.

The nest of buffering structures may have functional and evolutionary significances. The function of the buffering structure is analogically understood with a fire wall. It is a mechanism that keeps the effects of internal fluctuations inside and prevents them from being transmitted to the outside. In other words, the central metabolic pathway of *E. coli* has 17 nested firewall-like structures. Its characteristic property creates strong homeostasis or robustness outside against changes in enzyme amount and activity inside.

The following hypotheses may be possible to understand the biological significance of the mechanism to create external homeostasis in the central metabolic pathway in *E. coli*. The central metabolic pathway is a circuit in which chemical energy is extracted from nutrients (sugars) taken from the outside of a cell, and the activities or amounts of enzymes in the circuit have to be regulated always depending on the nutritional environment. On the other hand, in *E. coli* reaction system connecting to the central metabolic pathway, there are multiple subsystems that are indispensable for maintaining cell functions, such as amino acid synthesis or lipid synthesis pathways. Physiological problems will arise, if the regulation of enzyme activity in the central metabolic pathway influences these subsystems for maintaining cell functions. The presence of buffering structures may resolve this problem by keeping the effects of enzymatic regulation within the central metabolic pathway. For the same reason, buffering structures may exist at various parts in the chemical reaction network of organisms containing multiple subsystems with various physiological functions. Many buffering structures may be used in various parts of chemical reaction networks to achieve regulatory independence (modularity) between connected subsystems.

8.7 Perspective for Structural Sensitivity Analysis

As described in the previous section, buffering structures are defined from the topology of the reaction networks, and these structures have the property of absorbing the effects of internal fluctuations within it and not transmitting them to the outside under steady-state conditions. The law of localization may be a fundamental principle that creates homeostasis and adaptability in living systems. One of the characteristics of the behavior of living systems is a transient response in which a system reacts strongly to changes in the environment, such as light and nutrients, but returns to its original state with time. This behavior is called adaptation and is one of the important properties of life systems. To understand the transient response in adaptation, most of the previous studies have assumed a negative feedback from downstream to upstream in the signal transduction pathway reacting the environment. The law of localization (buffering structure) implies that adaptation can occur without such active regulation, simply by satisfying the topological condition in part of the network structure in the signal transduction pathway. Specifically, assume that part of the pathway is a buffering structure and that signals from the environment are input inside the buffering structure. In this case, the chemicals inside the buffering structure will continuously respond to changes in environmental signals, while the chemicals outside the buffering structure respond transiently, resulting in adaptation. Such a mechanism may be universal in living systems. In other words, a large proportion of the phenomena called adaptation or homeostasis in the life sciences may actually be understood by the law of localization.

The law of localization implies that the extent of regulation by an enzyme is determined solely by the local structure of the network. This may be used as a guideline for the stepwise identification of a large system from small substructures to large substructures. Figure 8.14 shows a hypothetical research strategy. First, candidate buffering structures are searched for based on the information in the database. Next, perturbations are actually applied to the inside of the candidate structures and the response of the system is measured. If, contrary to expectation, the perturbations affect the outside of the structure, it means that there is a mismatch between the

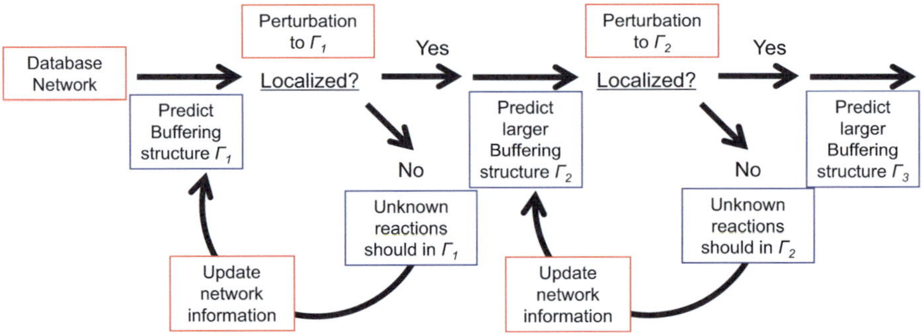

Fig. 8.14 A step-by-step strategy for system exploration. (From Okada and Mochizuki 2016)

database and the actual network. Since conditions for a buffering structure depend only on the local structure of network, the inconsistency should exist inside (or around) the candidate structure. We search for unknown reactions and regulations inside (or around) this structure, and update the information. By repeating this process of prediction and verification, it should be possible to determine the correct network step by step, from small structures to large ones.

Similarly, it can also be a guideline to determine the "extent of subsystem to be analyzed" based on the network structure. When studying the regulation of biological functions by enzymes, one should focus on the smallest buffering structure that contains the reaction of the enzyme of interest. This is because if the network information is correct, it is mathematically guaranteed that the effects of changes are not transmitted outside the buffering structure.

8.8 Summary

In this chapter, we introduced the structure sensitivity analysis for chemical reaction systems, which are fundamental to biological functions. This theory enables us to determine changes in chemical concentrations and reaction rates caused by changes in the activity or amount of enzymes in a chemical reaction system from the network structure. This theory relies only on network information and does not include model assumptions. This makes it possible to provide strong predictions to experimental biology. The model-free nature of the theory is its strength as a theory, but it also creates strengths as a tool to promote experimental biology, which is empirical research. Currently, there are many studies that attempt to understand dynamics of biological systems from the information of biomolecular networks. Using this theory, we will be able to logically understand the properties of biological systems by directly connecting them to network structures.

Mathematical Method 8A (Structure Sensitivity Analysis)

1. Derivation of **A**-matrix

In a chemical reaction system, dynamical change in the concentration of chemicals are generally described by a system of differential equations:

$$\frac{du_m}{dt} = \sum_{j=1}^{J} \nu_{mj} r_j\left(k_j; \boldsymbol{u}\right) \qquad (8\text{A}.1)$$

Here u_m $(m = 1, \cdots, M)$ is the concentration of chemical species U_m $(m = 1, \cdots, M)$ in the system. $r_j(k_j; \boldsymbol{u})$ is the reaction rate of reaction j $(j = 1, \cdots, J)$, which depends on the reaction parameter k_j and chemical concentration $\boldsymbol{u}$. From the reaction equation for reaction j:

$$j:\sum_{m=1}^{M} y_m^j U_m \rightarrow \sum_{m=1}^{M} \bar{y}_m^j U_m \tag{8A.2}$$

ν_{mj} is defined using the stoichiometric coefficient of the reaction as follows:

$$\nu_{mj} := \bar{y}_m^j - y_m^j$$

The $M \times J$ matrix $\boldsymbol{\nu} = (\nu_{mj})$ is termed a stoichiometric matrix. A stoichiometric matrix is usually expressed in terms of integer components.

In the following, we first consider the case where the dimension of the left null space $\ker(\boldsymbol{\nu}^T) = \{\boldsymbol{d} \in \mathbb{R}^M | \boldsymbol{d}^T\boldsymbol{\nu} = \boldsymbol{0}\}$ of $\boldsymbol{\nu}$ is zero $\dim(\ker(\boldsymbol{\nu}^T)) = 0$, that is, there exists no $\boldsymbol{d} \in \mathbb{R}^M$ satisfying $\boldsymbol{d}^T\boldsymbol{\nu} = \boldsymbol{0}$ except for the zero vector. In this case, $\boldsymbol{\nu}$ has full rank ($\mathrm{rank}(\boldsymbol{\nu}) = M \leq J$) from $\mathrm{rank}(\boldsymbol{\nu}) = M - \dim(\ker(\boldsymbol{\nu}^T)) = J - \dim(\ker(\boldsymbol{\nu}))$. This corresponds to the case where there is no conserved quantity in the dynamical system. If $\boldsymbol{d}$ is not a zero vector, then multiplying both sides of the differential equations by $\boldsymbol{d}^T$ from the left, we obtain:

$$\boldsymbol{d}^T \frac{d}{dt}\boldsymbol{u} = \boldsymbol{d}^T\boldsymbol{\nu}\boldsymbol{r} = \boldsymbol{0} \tag{8A.3}$$

which implies that $\boldsymbol{d}^T\boldsymbol{u} = \sum_{m=1}^{M} d_m u_m$ is the quantity conserved in the dynamics.

The reaction rate r_j is a function of the concentration of the substrate and the regulator chemicals. The specific form of the reaction rate function is unknown. Each reaction rate r_j also has a parameter k_j corresponding to the activity or amount of the enzyme.

Consider a steady state of this system (8A.1). In a steady state $(\widehat{\boldsymbol{r}}, \widehat{\boldsymbol{u}})$, total inflows and total outflows are balanced for each m, and we have:

$$\sum_{j=1}^{J} \nu_{mj}\widehat{r}_j\left(k_j; \widehat{\boldsymbol{u}}\right) = 0, \quad m = 1, \cdots, M \tag{8A.4}$$

Here, from the requirement that there exists a nonzero steady state reaction rate $\boldsymbol{r}$, $\dim(\ker(\boldsymbol{\nu})) > 0$ for null space of $\boldsymbol{\nu} = (\nu_{mj})$ (right null space, $\ker(\boldsymbol{\nu}) = \{\boldsymbol{c} \in \mathbb{R}^J | \boldsymbol{\nu}\,\boldsymbol{c} = \boldsymbol{0}\}$). The steady state reaction rate vector $\widehat{\boldsymbol{r}} = (\widehat{r}_1, \cdots, \widehat{r}_J)^T$ is represented by:

$$\widehat{\boldsymbol{r}} = \sum_{n=1}^{N} \mu^n \boldsymbol{c}^n \tag{8A.5}$$

where $\boldsymbol{c}^n(\ n = 1, \cdots, N)$ is the basis vector of the null space of the stoichiometric matrix. In the following discussion, we will consider the steady state $(\widehat{\boldsymbol{r}}, \widehat{\boldsymbol{u}})$ satisfying (8A.4) or (8A.5), but we will not write the hat $(\hat{\ })$ explicitly.

Since the rank of stoichiometric matrix ν is M, the dimension of the right null space is $N = \dim(\ker(\boldsymbol{\nu})) = J - M$. The basis vector $\boldsymbol{c}^n$ corresponds to the path of the steady flow determined from the network. A basis vector $\boldsymbol{c}^n$ can also be interpreted as a closed cycle created by the reactions in the network, or a path connecting from the inflow to the outflow. The definition of a cycle here is different from that in graph theory: a cycle consists of only reactions (edges) and does not include vertices. The components of $\boldsymbol{c}^n$ can be negative, so it is permissible to trace the reaction arrows in reverse in a cycle.

As can be seen from (8A.4), the steady state solution cannot be determined unless r_j is given specifically. However, even without determining a specific steady state solution, if we assume the existence of a steady state, it is possible to determine how the steady state varies with respect to the change in enzyme activity k_{j^*} $\left(k_{j^*} \to k_{j^*} + \delta k_{j^*}\right)$ from the network structure alone (Mochizuki and Fiedler 2015; Okada and Mochizuki 2016, 2017).

Assume that the steady state holds both before and after change in enzyme activity k_{j^*}. The change in reaction rate, $\delta_{j^*} r_j$ in response to change in activity $\left(k_{j^*} \to k_{j^*} + \delta k_{j^*}\right)$ is given in the following form using the total derivative:

$$\delta_{j^*} r_j = \left[\frac{\partial r_j}{\partial k_{j^*}} + \sum_{m=1}^{M} \frac{\partial r_j}{\partial u_m} \frac{\partial u_m}{\partial k_{j^*}}\right] \delta k_{j^*} \quad (j, j^* = 1, \cdots, J) \tag{8A.6}$$

Note that reaction rate $r_j(k_j; \boldsymbol{u})$ is a function of the reaction parameter k_j and chemical concentrations $\boldsymbol{u}$, and $\boldsymbol{u}$ may depend on k_{j^*}. Below, all partial derivatives are evaluated at the steady state.

(8A.6) implies that $\delta_{j^*} r_j$ is decomposed into a direct response to a change in k_{j^*} (first term), and an indirect response via a change in concentration u_m (second term). $\partial r_j / \partial k_{j^*}$ is nonzero (positive) only when $j = j^*$ and is zero when $j \neq j^*$. On the other hand, from (8A.5), the change in the reaction rate can be expressed as $\delta_{j^*} r_j = \sum_{n=1}^{N} \delta_{j^*} \mu^n c_j^n$, where $\delta_{j^*} \mu^n$ is the change in μ^n due to the change in enzymatic activity $\left(k_{j^*} \to k_{j^*} + \delta k_{j^*}\right)$. From the above, we obtain:

$$\frac{\partial r_j}{\partial k_{j^*}} \delta k_{j^*} + \sum_{m=1}^{M} \frac{\partial r_j}{\partial u_m} \frac{\partial u_m}{\partial k_{j^*}} \delta k_{j^*} = \sum_{n=1}^{N} \delta_{j^*} \mu^n c_j^n \quad (j, j^* = 1, \cdots, J) \tag{8A.7}$$

By a simple transformation, we have:

$$\sum_{m=1}^{M} r_{jm} \delta_{j^*} u_m - \sum_{n=1}^{N} \delta_{j^*} \mu^n c_j^n = -\frac{\partial r_j}{\partial k_{j^*}} \delta k_{j^*} \quad (j, j^* = 1, \cdots, J) \tag{8A.8}$$

Here, we set $r_{jm} := \frac{\partial r_j}{\partial u_m}$ and $\delta_{j^*} u_m := \frac{\partial u_m}{\partial k_{j^*}} \delta k_{j^*}$. For an easier understanding, we show another form:

$$\begin{pmatrix} r_{j1} & \cdots & r_{jm} & \cdots & r_{jM} \end{pmatrix} \begin{pmatrix} \delta_{j^*} u_1 \\ \vdots \\ \delta_{j^*} u_m \\ \vdots \\ \delta_{j^*} u_M \end{pmatrix} + \begin{pmatrix} -c_j^1 & \cdots & -c_j^n & \cdots & -c_j^N \end{pmatrix}$$

$$\times \begin{pmatrix} \delta_{j^*} \mu^1 \\ \vdots \\ \delta_{j^*} \mu^n \\ \vdots \\ \delta_{j^*} \mu^N \end{pmatrix} = \begin{cases} -\dfrac{\partial r_j}{\partial k_{j^*}} \delta k_{j^*} & (j = j^*) \\ 0 \quad (j \neq j^*) \end{cases} \quad (j, j^* = 1, \cdots, J) \tag{8A.9}$$

From (8A.8) or (8A.9) by taking j in row and j^* in column, we have:

$$\left(\begin{array}{ccc|ccc} r_{11} & \cdots & r_{1M} & -c_1^1 & \cdots & -c_1^N \\ \vdots & & \vdots & \vdots & & \vdots \\ r_{J1} & \cdots & r_{JM} & -c_J^1 & \cdots & -c_J^N \end{array} \right) \left(\begin{array}{ccc} \delta_1 u_1 & \cdots & \delta_J u_1 \\ \vdots & & \vdots \\ \delta_1 u_M & \cdots & \delta_J u_M \\ \hline \delta_1 \mu^1 & \cdots & \delta_J \mu^1 \\ \vdots & & \vdots \\ \delta_1 \mu^N & \cdots & \delta_J \mu^N \end{array} \right) =$$

$$-\begin{pmatrix} \dfrac{\partial r_1}{\partial k_1} \delta k_1 & & 0 \\ & \ddots & \\ 0 & & \dfrac{\partial r_J}{\partial k_J} \delta k_J \end{pmatrix} \tag{8A.10}$$

Let $\mathbf{A} := (r_{jm} | -\boldsymbol{c}^n)$ be the first matrix on the left-hand side of Eq. (8A.10). The right-hand side is a diagonal matrix, where only the diagonal components ($j = j^*$) take nonzero value. Since we are only interested in the qualitative response here, the right-hand side can be taken as an identity matrix. Assuming that $\mathbf{A}$ has an inverse matrix, we obtain:

$$\mathbf{S} \equiv \begin{pmatrix} \delta_1 u_1 & \cdots & \delta_J u_1 \\ \vdots & & \vdots \\ \delta_1 u_M & \cdots & \delta_J u_M \\ \hline \delta_1 \mu^1 & \cdots & \delta_J \mu^1 \\ \vdots & & \vdots \\ \delta_1 \mu^N & \cdots & \delta_J \mu^N \end{pmatrix} \propto -\mathbf{A}^{-1} \tag{8A.11}$$

In Eq. (8A.11), each column identifies perturbed reaction parameter ($j^* = 1, \cdots, J$) and each row identifies molecular species ($m = 1, \cdots, M$) or basis vector ($n = 1, \cdots, N$) of $\boldsymbol{\nu}$. Each component of $\mathbf{S}$ shows the response of concentration u_m of each chemical m (or that of steady flow μ^n of cycle n) to the change in enzyme activity k_{j^*} of reaction j^*. Eq. (8A.11) gives the response to an increase in the activity (e.g., overexpression) of each enzyme. If one is interested in the response to a decrease in activity (e.g., knockdown) k_{j^*}, the minus sign on the right-hand side should be simply removed.

2. Structural determination of sensitivity of a reaction network

The above explanation shows the derivation of the formula. To analyze the perturbation responses from the network structure in practice, one of the following criteria is used depending on the case.

(i) Determining the presence or absence of a response, i.e., zero or nonzero of $\delta_{j^*} u_m$ or $\delta_{j^*} \mu^n$.

Zero or nonzero of each component of the matrix $\mathbf{A}$ is determined solely from the structure of the reaction network, i.e. stoichiometric matrix and dependence of reaction rate functions. That is, r_{jm} is constitutively zero when reaction j does not depend on the chemical species m, and nonzero when it does (unless special values of concentrations are chosen). The components of $\boldsymbol{c}^n$ are also determined from the stoichiometric matrix. In other words, the pattern of the zero (nonzero) components of matrix $\mathbf{A}$ reflects the structure of the reaction network.

Zero or nonzero of each component S_{ij} in the matrix $\mathbf{S}$ reflects the presence or absence of change in each concentration or each reaction rate to perturbation of each reaction parameter. They can be determined solely from the pattern of the zero (nonzero) components of matrix $\mathbf{A}$, i.e., the structure of the reaction network.

The Eq. (8A.7) uses the partial derivative, it appears to hold only when the change δk_{j^*} in enzymatic activity k_{j^*} is small. However, the pattern of zero components of matrix $\mathbf{A}$ is determined from the network structure alone and does not depend on parameter $\boldsymbol{k}$ or the concentration $\boldsymbol{u}$. Thus, the pattern of zero (nonzero) components calculated by $\mathbf{A}^{-1}$ (the response pattern in matrix $\mathbf{S}$) will be always the same, even after repeated application of the change in k_{j^*}, $\left(k_{j^*} \rightarrow k_{j^*} + \delta k_{j^*}\right)$. In other words, the response pattern does not change, even if the change δk_{j^*} in parameter is large.

(ii) Determining the direction of a response, i.e., sign of $\delta_{j^*} u_m$ or $\delta_{j^*} \mu^n$.

Suppose we want to determine not only the presence or absence of a response but also direction of responses, i.e. whether the concentration or reaction rate increases or decreases. In this case, we need to assume that the reaction rates r_j are monotonic functions of the concentration of the influencing chemicals. For example, assume that it is an increasing function of substrate chemicals and an increasing (positive regulation) or decreasing (negative regulation) function of regulator chemicals. As a result, r_{jm} is determined to be zero, positive, or negative. We usually indicate a negative component explicitly by writing $-r_{jm}$ ($r_{jm} > 0$). The matrix **A** with sign information in addition to zero or nonzero information of each component is given. Then we calculate the inverse of the matrix **A** to determine the sign of $\delta_{j^*} u_m$ and $\delta_{j^*} \mu^n$.

Empirically, unless the structure of the network is too complex, it is possible to determine the sign of the response S_{ij} uniquely in many cases. When networks are too complicated, (though zero or nonzero of components S_{ij} are always determined uniquely) it may happen that the signs of some nonzero components are not uniquely determined. This is because the components in the inverse matrix **S** contain multiple r_{jm}s and their signs may be different. In this case, the signs of such responses can be determined again by giving the relative magnitudes of some r_{jm}s contained in the components.

3. Structural Sensitivity Analysis for a system with conserved quantities.

As mentioned at the beginning, so far we have assumed that the dimension of the left null space of $\boldsymbol{\nu}$ is 0, that is, there is no conserved quantity in the system. If there are conserved quantities in the system, in addition to (8A.8), we have another set of equations

$$\sum_{m=1}^{M} d^l_m u_m = T^l \quad (l = 1, \cdots, L). \tag{8A.12}$$

Here, $\boldsymbol{d}^l$ ($l = 1, \cdots, L$) are the basis vectors of the left null space ker $(\boldsymbol{\nu}^T) = \{\boldsymbol{d} \in \mathbb{R}^M | \boldsymbol{d}^T \boldsymbol{\nu} = \boldsymbol{0}\}$ of the stoichiometric matrix $\boldsymbol{\nu}$, and L is the dimension of this space $L =$ dim (ker($\boldsymbol{\nu}^T$)). The values of conserved quantities T^l are given by the initial condition of dynamics (8A.1). By combining (8A.8) and (8A.12), we obtain the extended matrix:

$$\mathbf{A} := \left(\begin{array}{ccc|ccc} r_{11} & \cdots & r_{1M} & & & \\ \vdots & & \vdots & -\boldsymbol{c}^1 & \cdots & -\boldsymbol{c}^N \\ r_{J1} & \cdots & r_{JM} & & & \\ \hline & -(\boldsymbol{d}^1)^T & & 0 & \cdots & 0 \\ & \vdots & & \vdots & & \vdots \\ & -(\boldsymbol{d}^L)^T & & 0 & \cdots & 0 \end{array}\right). \tag{8A.13}$$

By calculating the inverse of this matrix $\mathbf{A}$, the response of the system to the perturbation of parameters is determined at once:

$$\mathbf{S} \equiv \left(\begin{array}{ccc|ccc} \delta_1 u_1 & \cdots & \delta_J u_1 & \delta_{J+1} u_1 & \cdots & \delta_{J+L} u_1 \\ \vdots & & \vdots & \vdots & & \vdots \\ \delta_1 u_M & \cdots & \delta_J u_M & \delta_{J+1} u_M & \cdots & \delta_{J+L} u_M \\ \hline \delta_1 \mu^1 & \cdots & \delta_J \mu^1 & \delta_{J+1} \mu^1 & \cdots & \delta_{J+L} \mu^1 \\ \vdots & & \vdots & \vdots & & \vdots \\ \delta_1 \mu^N & \cdots & \delta_J \mu^N & \delta_{J+1} \mu^N & \cdots & \delta_{J+L} \mu^N \end{array}\right) \propto -\mathbf{A}^{-1} \tag{8A.14}$$

Here, $\delta_{J+l} u_m$ is the change in the steady state concentration of chemical m when the conserved quantity T^l corresponding to the lth basis vector $\boldsymbol{d}^l$ of the left null space of ν is perturbed. The $\delta_{J+l} \mu^n$ is the change in the steady flow along the basis vector $\boldsymbol{c}^n$ in the null space of ν when T^l is perturbed. For details of the analysis, see Okada and Mochizuki (2017).

Mathematical Method 8B (Law of Localization)

For the chemical reaction system and the matrix $\mathbf{A}$ discussed in the 8A mathematical method, the law of localization and the concept of buffering structures are given as follows (Okada and Mochizuki 2016, 2017; Yamauchi et al. 2024).

1. Preparation

Consider a chemical reaction network $\Gamma = (\mathcal{M}, \mathcal{J})$ and an arbitrary subgraph $\Gamma^* = (\mathcal{M}^*, \mathcal{J}^*)$ of Γ. Here, $\mathcal{M}^* \subseteq \mathcal{M}$ is a subset of chemicals and $\mathcal{J}^* \subseteq \mathcal{J}$ is a subset of reactions. For the stoichiometric matrix $\boldsymbol{\nu}$ of Γ, consider a part of the matrix $\boldsymbol{\nu}_{\mathcal{J}^*} := \left(\nu_{m,j}\right)_{m \in \mathcal{M}, j \in \mathcal{J}^*}$ that only contains the columns for $j \in \mathcal{J}^*$, and let $N^* := \dim(\ker(\boldsymbol{\nu}_{\mathcal{J}^*}))$. N^* is the number of basis vectors in the right null space of $\boldsymbol{\nu}_{\mathcal{J}^*}$, i.e., the number of independent cycles consisting of only reactions in $\mathcal{J}^*$. Similarly, consider the partial stoichiometric matrix $\boldsymbol{\nu}_{\mathcal{M}^*} := \left(\nu_{mj}\right)_{m \in \mathcal{M}^*, j \in \mathcal{J}}$ that only contains the rows for $m \in \mathcal{M}^*$. $\dim(\operatorname{coker}(\boldsymbol{\nu}_{\mathcal{M}^*})) \left(= \dim\left(\ker\left(\boldsymbol{\nu}_{\mathcal{M}^*}^T\right)\right)\right)$ is the number of basis vectors of the left null space of $\boldsymbol{\nu}_{\mathcal{M}^*}$, that is, the number of independent conserved quantities consisting of only the chemicals contained in $\mathcal{M}^*$. However, what is important here is $L^* := \dim(\operatorname{coker}(\boldsymbol{\nu})) - \dim(\operatorname{coker}(\boldsymbol{\nu}_{\mathcal{M} \setminus \mathcal{M}^*}))$, i.e., "the number of independent conserved quantities that contain at least one chemical in $\mathcal{M}^*$."

2. Definition (Buffering Structure)

A subgraph $\Gamma^* = (\mathcal{M}^*, \mathcal{J}^*)$ of a chemical reaction network $\Gamma = (\mathcal{M}, \mathcal{J})$ is called "output complete" if it satisfies the following condition:

$$\mathcal{J}^* \supseteq \left\{ j, \middle| , \frac{\partial r_j}{\partial u_m} \neq 0, m \in \mathcal{M}^* \right\}. \tag{8B.1}$$

For an output-complete subgraph Γ^*, the following index χ is defined as:

$$\chi(\Gamma^*) := -|\mathcal{M}^*| + |\mathcal{J}^*| - N^* + L^*. \tag{8B.2}$$

An output-complete subgraph Γ^* is called a "buffering structure" if it satisfies:

$$\chi(\Gamma^*) = 0. \tag{8B.3}$$

Note 1: Output completeness implies that "all reactions influenced by chemicals in $\mathcal{M}^*$ are included in $\mathcal{J}^*$." In other words, once a subset $\mathcal{M}^*$ of chemicals is chosen, all reactions that use them as substrates or regulators are included in the subset $\mathcal{J}^*$ of reactions. After satisfying this condition, any additional reactions may be added to $\mathcal{J}^*$.

Note 2: $\chi(\Gamma^*)$ means "– (number of chemicals) + (number of reactions) – (number of cycles) + (number of conserved quantities)" of Γ^*. When there is no conserved quantity ($L^* = 0$), it corresponds to the alternating sum of the number of zero-, one-, and two-dimensional structures contained in the subgraph Γ^*. It is thought that this index χ has a strong relationship with the Euler characteristic. When the **A** matrix is regular, the value of $\chi(\Gamma^*)$ for an output complete graph Γ^* is generally nonnegative (either 0 or positive).

3. Theorem (Law of Localization)

   Consider a steady state of a chemical reaction system. When a reaction parameter within a buffering structure (or a conserved quantity containing at least one chemical in a buffering structure) is perturbed, the steady state concentration of chemicals outside the buffering structure and the steady state rate of reactions outside the buffering structure do not change. In other words, the response of the system to the perturbation remains within the buffering structure.

Proof
When a buffering structure exists in a chemical reaction system, by swapping the rows and columns of matrix **A**, we can construct a block matrix with the elements related to $\Gamma^* = (\mathcal{M}^*, \mathcal{J}^*)$ gathered in the upper left corner.

$$\mathbf{A} = \begin{pmatrix} \boxed{\mathbf{A}_{\Gamma^*}} & * \\ \mathbf{0} & \boxed{\mathbf{A}_{\overline{\Gamma^*}}} \end{pmatrix} \tag{8B.4}$$

Here, $\boxed{\mathbf{A}_{\Gamma^*}}$ is a square nonzero block of size $(|\mathcal{J}^*| + L^*) \times (|\mathcal{M}^*| + N^*)$, where $|\mathcal{J}^*| + L^* = |\mathcal{M}^*| + N^*$. Each row of $\boxed{\mathbf{A}_{\Gamma^*}}$ corresponds to a reaction in $\mathcal{J}^*$ or a

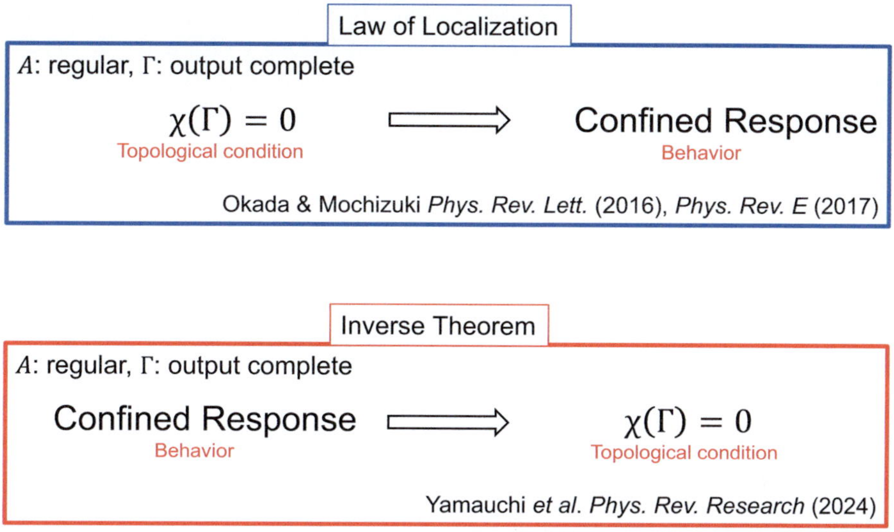

Fig. 8.B1 The relationship between the law of localization and the inverse theorem

conserved quantity that contains at least one chemical in $\mathcal{M}^*$, each column corresponds to a chemical in $\mathcal{M}^*$ or an independent cycle in $\mathcal{J}^*$. $\boxed{\mathbf{A}_{\Gamma^*}}$ and $\boxed{\mathbf{A}_{\overline{\Gamma^*}}}$ are nonzero square blocks, and when $\mathbf{A}$ is invertible (regular), they are also invertible. The $\mathbf{0}$ is a block of size $(|\mathcal{J}| - |\mathcal{J}^*| - L^*) \times (|\mathcal{M}^*| + N^*)$ (not necessarily square) with all components being 0. From the pattern of the zero components in the matrix $\mathbf{A}$, the inverse matrix has the same pattern of zero components, i.e.,

$$\mathbf{S} \propto -\mathbf{A}^{-1} = \begin{pmatrix} \boxed{-\mathbf{A}_{\Gamma^*}^{-1}} & * \\ \mathbf{0} & \boxed{-\mathbf{A}_{\overline{\Gamma^*}}^{-1}} \end{pmatrix}. \tag{8B.4}$$

Again, $\mathbf{0}$ is a block of size $(|\mathcal{J}| - |\mathcal{M}^*| - N^*) \times (|\mathcal{J}^*| + L^*)$ where all components are 0. Each component of $\mathbf{0}$ indicates the change in concentration of an external chemical ($m \notin \mathcal{M}^*$), or the change in an external steady-state flow when a parameter of a reaction $j^* \in \mathcal{J}^*$ inside the buffering structure (or a conserved quantity associated to the buffering structure) is perturbed. Thus, when a chemical reaction system contains a buffering structure, the responses to a perturbation of a reaction parameter inside the buffering structure (or that of a conserved quantity associated to the buffering structure) are confined to the inside of the buffering structure.

4. Inverse Theorem (Fig. 8.B1)

Yamauchi et al. (2024) proved that the inverse of the law of localization holds. In other words, if the effects of a parameter perturbation are localized to a finite range of a network, then there is always a corresponding buffering structure. This indicates that there is a one-to-one relationship between the topology of a network and the behavior of the system.

Exercise

Exercise 8.1

For the Eq. (8.2), find the equilibrium point $(\widehat{u}_A, \widehat{u}_B)$ by substituting specific functions $r_1 := c$, $r_2 := au_A$, $r_3 := bu_B$. Consider a perturbation to the reaction parameter of reaction 2 by increasing the parameter a. How does the equilibrium point change to the perturbation?

Exercise 8.2

As in Exercise 8.1, for the system of differential equations with $r_1 := c$, $r_2 := au_A$, and $r_3 := bu_B$, find the time development solutions of the concentrations $u_A(t)$ and $u_B(t)$. A differential equation that has a linear term such as $du/dt = c - au$, as well as a constant term (or a term that only depends on time), is called an inhomogeneous linear differential equation. In this case, the solution is obtained by substituting $u(t) := A(t)e^{-at}$ into this differential equation and solving the identity equation for $A(t)$.

Exercise 8.3

For the dynamics (8.2) of the reaction system in Fig. 8.3, find the basis of the null space of the stoichiometric matrix. Using this, show that the matrix **A** is as in Eq. (8.5). Also, find the inverse matrix of this matrix **A** and show that the sensitivity matrix **S** is given as in Eq. (8.8).

Exercise 8.4

For the dynamics (8.9) of the reaction system in Fig. 8.6, find the basis of the null space of the stoichiometric matrix. Using this, show that matrix **A** is as in Eq. (8.10). Also, find the inverse matrix of this matrix **A** and show that the sensitivity matrix **S** is given as in Eq. (8.11).

Exercise 8.5

Find all the buffering structures in the reaction system shown in Fig. 8.6. Confirm that these explain the response pattern in (8.11) or Fig. 8.7.

References

Clarke BL (1980) Stability of complex reaction networks. In: Advances in chemical physics 43. Wiley, New York, pp 1–215

Feinberg M (2019) Foundations of chemical reaction network theory, Applied Mathematical Sciences (202). Springer, Cham, p 483

Ferjani A, Kawade K, Asaoka M, Oikawa A, Okada T, Mochizuki A, Maeshima M, Hirai MY, Saito K, Tsukaya H (2018) Pyrophosphate inhibits gluconeogenesis by restricting UDP-glucose formation in vivo. Sci Rep 8:14696

Ishii N et al (2007) Multiple high-throughput analyses monitor the response of E. coli to perturbations. Science 316:593–597

Mochizuki A, Fiedler B (2015) Sensitivity of chemical reaction networks: a structural approach. 1. Examples and the carbon metabolic network. J Theor Biol 367:189–202

Okada T, Mochizuki A (2016) Law of localization in chemical reaction networks. Phys Rev Lett 117:048101

Okada T, Mochizuki A (2017) Sensitivity and network topology in chemical reaction systems. Phys Rev E 96:022322

Yamauchi Y, Hishida A, Okada T, Mochizuki A (2024) Finding regulatory modules of chemical reaction systems. Phys Rev Res 6:023150

Chapter 9
Chemical Reaction Systems II: Behavioral Diversity and Bifurcation

Abstract The behavior of biological systems may change discontinuously in response to continuous changes in the environment or external signals. In this chapter, we will understand this phenomenon as bifurcation behavior in the dynamics of biological systems. The importance of bifurcation in biological functions and their understanding is gradually increasing. This chapter is also closely related to the regulation of chemical reaction systems studied in the previous chapter. In Chap. 8, it was shown that the response range of a chemical reaction system to an enzymatic change is determined by the "buffering structure." In this chapter, we show that buffering structures govern not only enzymatic regulation but also another important aspect: bifurcation.

Keywords Bifurcation · Hysteresis · Structural bifurcation analysis · Factorization of condition · Inducing parameter · Bifurcating chemical

9.1 Bifurcations in Living Systems

In Chaps. 3 and 4, we discussed the concept of bifurcation, a phenomenon in which the solution of a dynamical system exhibits a qualitative change in response to a continuous change in its parameter values. Bifurcation has captured the interest of mathematicians, but it is not limited to that. In actual biological phenomena, there are known examples of bifurcation that have been observed experimentally and that play an important role in biological functions.

In *E. coli*, lactose permease (LacY) is responsible for the cellular uptake of lactose and its analogs (TMG, etc.). Ozbudak et al. found an interesting phenomenon in which the expression level of LacY switches depending on the nutrient environment (Fig. 9.1b) (Ozbudak et al. 2004). They conducted an experiment in which the concentration of thio-methylgalactoside (TMG), a lactose analog, was varied while *E. coli* were cultured successively, i.e., a population of *E. coli* experienced a change in nutritional environment under culture. When the TMG concentration was increased from low to high, the expression of an uptake protein (LacY) in the population increased rapidly at a certain concentration (around 30 μM) (Fig. 9.1b, bottom). On the other hand, when the TMG concentration was decreased from high

A. Mochizuki, *Structural Theories for Dynamical Networks in Biology*, Theoretical Biology, https://doi.org/10.1007/978-981-96-6026-1_9

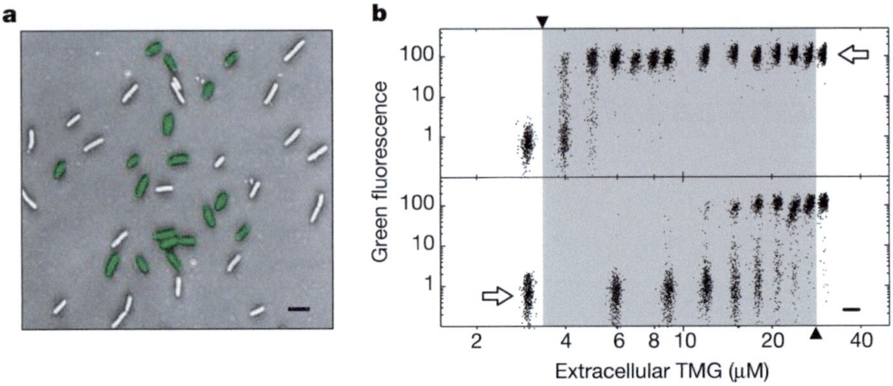

Fig. 9.1 Nutrient environment-dependent lactose uptake by *E. coli*. (**a**) *E. coli* in culture and GFP fluorescence corresponding to LacY expression. The gene of green fluorescent protein (GFP) was transfected into *E. coli* under the *lacY* promoter. (**b**) Expression levels of LacY at different nutrient concentrations. Vertical axis: LacY expression captured by GFP fluorescence; horizontal axis: Lactose analog (TMG) concentration. (Ozbudak et al. 2004)

to low, the expression level of LacY dropped sharply at another threshold concentration (Fig. 9.1b top), which was different and lower from the threshold concentration observed when the TMG concentration was increased.

It is known that lactases, such as TMG, activate the expression of the LacY gene after being taken up into the cell (Fig. 9.2a). There is a positive feedback such that an increase in the intracellular TMG concentration promotes transcription of LacY, which in turn increases the TMG concentration. Ozbudak et al. developed a simple mathematical model to analyze this system. A somewhat simplified version of Ozbudak's model is:

$$\frac{du}{dt} = \beta v - u, \tag{9.1a}$$

$$\frac{dv}{dt} = \alpha \frac{1 + u^n}{\rho + u^n} - v. \tag{9.1b}$$

where u is the TMG concentration in the cell, v is the expression level of LacY. β is a parameter representing the TMG concentration in the extracellular environment, and ρ (>1) is a parameter. The first term on the right-hand side of (9.1b) is a function that increases in an S-shape with respect to u. A result of the nullcline analysis of this model is shown in Fig. 9.2b. It is seen that the number of nullcline intersections, or equilibrium points, changes from one to three depending on the extracellular TMG concentration β. More directly, Fig. 9.3 shows these equilibrium points (LacY expression level v) as a function of the parameter β.

The number and value of equilibrium points change depending on the value of the parameter β. For a small value of β, the model has only one stable equilibrium point with a low value of LacY expression. When the extracellular TMG concentration β is

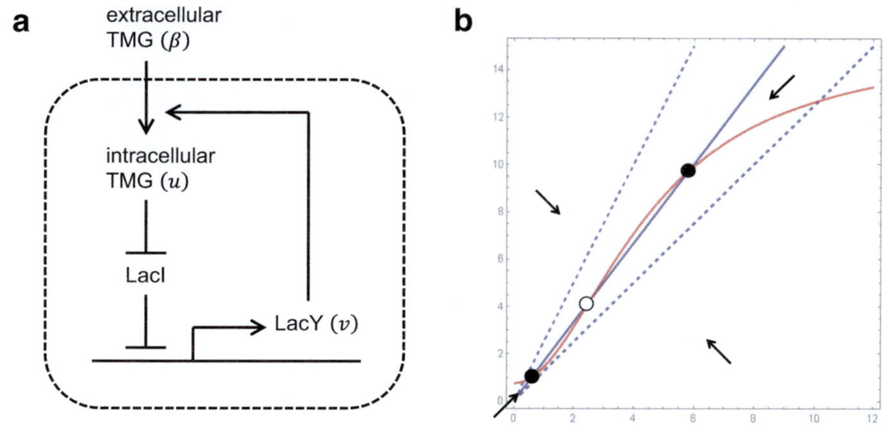

Fig. 9.2 Mathematical model for lactose uptake by *E. coli* and its analysis. (**a**) Regulatory relationship between lactose analog (TMG) concentration and LacY (Lactose uptake enzyme) expression. (**b**) Nullcline analysis of the model (9.1). The horizontal axis is the intracellular TMG concentration (u) and the vertical axis is the LacY expression level (v). The blue and red lines indicate the nullclines of the intracellular TMG concentration (u) and the expression level of the uptake protein LacY (v), respectively. Black and white circles indicate stable and unstable equilibrium points, respectively. Arrows indicate the direction of change over time. Parameter values are $\alpha = 15$, $\beta = 0.6$, $\rho = 20$, and $n = 2$. Dashed lines indicate nullclines of intracellular TMG concentration (u) when $\beta = 0.4$ and $\beta = 0.8$

Fig. 9.3 Equilibrium points of the model (9.1) determined by numerical analysis. Nutrient uptake gene LacY expression levels for various nutrient environments are shown. Red: stable equilibrium points, blue: unstable equilibrium point

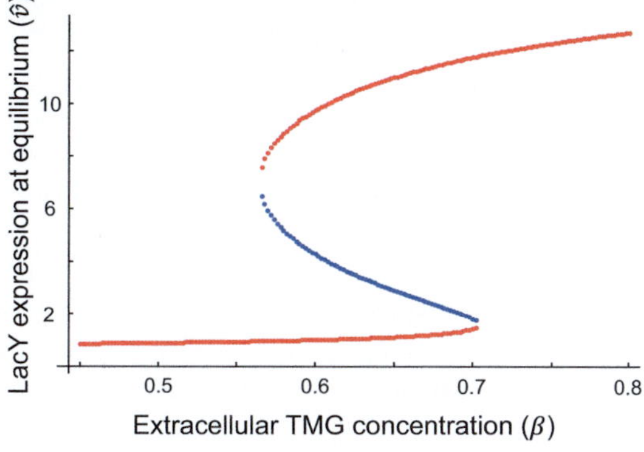

increased, at a certain value of β, a pair of two equilibria with larger LacY expressions appear. One of the two equilibrium points is stable (stable node) and the other is unstable (saddle point). Such a qualitative change in the solution is called a bifurcation, and the point where the bifurcation occurs on the parameter β axis is called the bifurcation point. A bifurcation where the stable node and the saddle point appear or disappear simultaneously as a pair is called a saddle-node bifurcation. When the extracellular TMG concentration β is further increased, the unstable equilibrium point and the originally existing stable equilibrium point with low

LacY expression merge, and the pair of the equilibria disappears. That is, another saddle-node bifurcation occurs here.

From Fig. 9.3, we can understand why the sudden increase or decrease of LacY expression occurred, and why the TMG concentrations β at which the sudden changes occurred were different between the cases of increasing and decreasing the TMG concentration. When TMG was changed from low to high values, the system state moved along the branch of the stable equilibrium point with low level of LacY, and at the bifurcation point where this stable equilibrium is lost, it moved to another stable equilibrium point with high expression of LacY. On the other hand, when TMG was changed from a high value to a low value, the system state moved along the branch of the stable equilibrium point with a high expression of LacY and jumped to another stable equilibrium at the bifurcation point where this stable equilibrium is lost. This difference in behavior depending on the state of the system (in this case, on which branch it exists) is called the hysteresis phenomenon. The hysteresis is sometimes interpreted as a mechanism for memorizing the environment.

The behavior of LacY expression observed in Fig. 9.1b can be understood as a hysteresis phenomenon of this system. In other words, it is expected that the real *E. coli* system has bistable solutions depending on the TMG concentration and two saddle-node bifurcation points within the range of manipulated TMG concentrations. The complex behavior of the system can be understood through the analysis of the mathematical model, which also demonstrates the validity of the model.

The cell cycle is another example where bifurcation plays an important role in biological phenomena. Cell division is, of course, the most fundamental process of life. From one division to the next, four phases must be passed through: G1, S, G2, and M. Collectively, these are called the cell cycle. Genetics and molecular biology have revealed that the cell cycle is a complex system that is regulated by the activity of several species of proteins such as Cdk and cyclins and includes many positive and negative feedbacks between them (Fig. 9.4).

Novak and Tyson constructed and analyzed a mathematical model of the cell cycle system (Novak and Tyson 1993). They found that the essence of cell cycle dynamics is not periodic oscillations but rather irreversible transitions between different steady states of the system. The amount of cyclin that changes with time plays the role of a bifurcation parameter, and the transitions between the different phases may correspond to transitions of the steady state at the bifurcation point.

9.2 Eigenvalues at Bifurcation Points

Let us proceed with a further mathematical analysis of the bifurcation point (see Fig. 9.3) in the model (9.1). From the nullcline analysis in Fig. 9.2b, the bifurcation point can be identified as the tangent point of two nullclines (the equilibrium points are the double roots). From the conditions of equilibrium of (9.1) and the condition of the double roots, $\widehat{u}$ at the point $(\widehat{u}, \widehat{v})$ satisfies the following two equations:

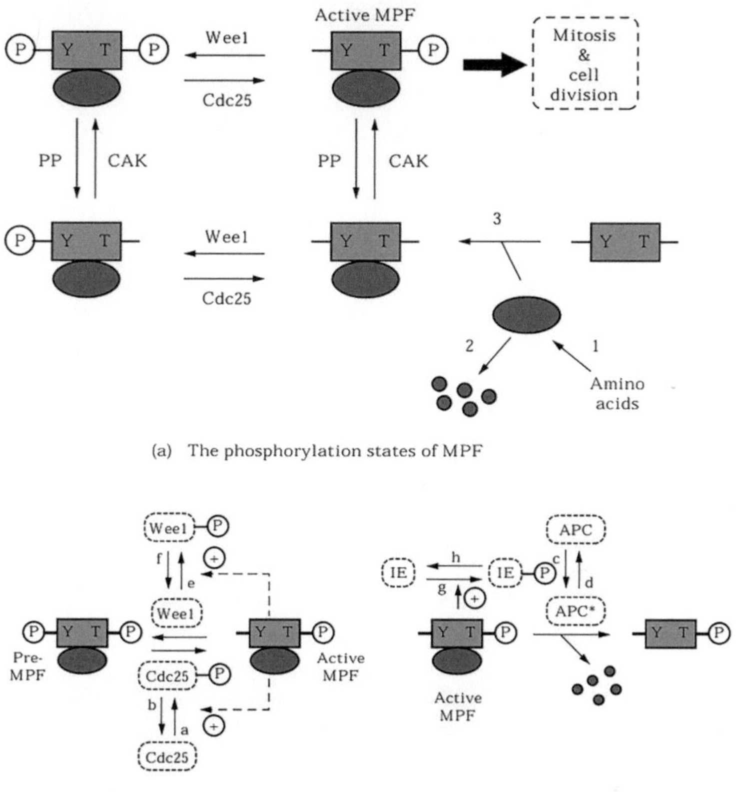

(a) The phosphorylation states of MPF

(b) Positive feedback loops

(c) Negative feedback loop

Fig. 9.4 Regulatory network of molecules involved in the cell cycle of the African clawed frog (Borisuk and Tyson 1998). Rectangles indicate CDK1 molecules and ovals indicate CyclinB molecules

$$\alpha\beta\frac{1+\widehat{u}^n}{\rho+\widehat{u}^n}-\widehat{u}=0, \tag{9.2a}$$

$$\alpha\beta\frac{(\rho-1)n\widehat{u}^{n-1}}{(\rho+\widehat{u}^n)^2}-1=0. \tag{9.2b}$$

We now perform a local stability analysis for the equilibrium point $(\widehat{u},\widehat{v})$. The stability of the equilibrium point is determined by the sign of the largest real part of the eigenvalues λ of the Jacobi matrix $\boldsymbol{J}$, and the eigenvalues λ are given by solving the eigenequation:

$$\det(\lambda\boldsymbol{E}-\boldsymbol{J})=0. \tag{9.3}$$

The Jacobian matrix near the equilibrium point $(\widehat{u}, \widehat{v})$ in model (9.1) is given by

$$J = \begin{pmatrix} -1 & \beta \\ \alpha \dfrac{(\rho-1)n\widehat{u}^{n-1}}{(\rho+\widehat{u}^n)^2} & -1 \end{pmatrix}. \tag{9.4}$$

Equation (9.2b) implies that the determinant of the Jacobian matrix det($\boldsymbol{J}$) is zero. The eigenequation is given as $\lambda(\lambda + 2) = 0$ and the maximum eigenvalue λ is 0. In general, when an equilibrium point of a system is the double root, the determinant det($\boldsymbol{J}$) of the Jacobi matrix $\boldsymbol{J}$ is zero and there exists an eigenvalue of $\lambda = 0$.

In general, when a bifurcation of the equilibrium point(s) occurs at a certain parameter value $k = k^*$, there is a real eigenvalue with $\lambda(k^*) = 0$ and det($\boldsymbol{J}(k^*)$) = 0 at that point $k = k^*$. Series of types of bifurcations involving equilibrium states, such as saddle-node bifurcations, pitch-fork bifurcations, and transcritical bifurcations, are captured by the criterion det($\boldsymbol{J}(k^*)$) = 0. On the other hand, there are other types of bifurcations such as Hopf bifurcations, which are caused by the sign of the real part of the complex eigenvalue being zero. Hopf bifurcations of two-dimensional systems can be captured not by det($\boldsymbol{J}$) = 0, but by tr($\boldsymbol{J}$) = 0.

Based on the above argument, some mathematical studies have discussed the conditions for a bifurcation of equilibrium points based only on the components of the Jacobian matrix of the system of interest. However, as the number of variables increases, the size of the matrix also increases, and this method does not always make clear conclusions.

9.3 Structural Bifurcation Analysis of Chemical Reaction Systems

Up to this point, we have discussed the bifurcation of systems without distinguishing whether the system of interest is a regulatory or a reaction system. From here on, we will confine our targets to reaction systems described by sets of state transitions. For reaction systems, it is possible to discuss bifurcation behaviors directly in relation to network structure. We call the analytical method "structural bifurcation analysis." The "buffering structure" discussed in Chap. 8 plays an important role in a different sense here. First, let us summarize the results of Chap. 8 (Mochizuki and Fiedler 2015; Okada and Mochizuki 2016, 2017). For simplicity, we will explain the case where there is no conserved quantity in the system. See 8A and 8B Mathematical Method for the details.

For the steady state of a system of differential equations describing the dynamics of a chemical reaction system,

$$\frac{du_m}{dt} = \sum_{j=1}^{J} \nu_{mj} r_j(k_j; \boldsymbol{u}) = 0, \quad m = 1, \cdots, M. \tag{9.5}$$

The change in each chemical concentration u_m (or reaction rate r_j) for a perturbation of each reaction parameter k_j (a parameter associated with each reaction, i.e., $\partial r_{j^*}/\partial k_{j^*} > 0$) is given by

$$\mathbf{S} \equiv \begin{pmatrix} \delta_1 u_1 & \cdots & \delta_J u_1 \\ \vdots & & \vdots \\ \delta_1 u_M & \cdots & \delta_J u_M \\ \hline \delta_1 \mu^1 & \cdots & \delta_J \mu^1 \\ \vdots & & \vdots \\ \delta_1 \mu^N & \cdots & \delta_J \mu^N \end{pmatrix} \propto -\mathbf{A}^{-1}, \tag{9.6a}$$

$$\mathbf{A} = \left(\frac{\partial r_j}{\partial u_m} \middle| -\boldsymbol{c}^n \right). \tag{9.6b}$$

where $\delta_j u_m$ denotes the change in concentration of chemical m when the parameter of reaction j is perturbed (8A Mathematical Method). $\delta_j \mu^n$ denotes the change in the coefficient of the steady-state flow along $\boldsymbol{c}^n$ when the parameter of reaction j is perturbed. The left part of matrix $\mathbf{A}$, $\partial r_j/\partial u_m$, is the partial derivative of the reaction rate r_j with respect to chemical concentration u_m, where the rows correspond to each reaction $j = 1, \cdots, J$ and the columns correspond to chemical species $m = 1, \cdots, M$. $\partial r_j/\partial u_m$ takes nonzero values when reaction j depends on chemical m and zero when it does not. The right part $\boldsymbol{c}^n$ of matrix $\mathbf{A}$ is the basis vector in null space of the stoichiometric matrix $\boldsymbol{\nu}$, corresponding to the steady-state flow path determined by the topology of the network. Pattern of zero or nonzero of each component of matrix $\mathbf{A}$ directly reflects the network structure.

The buffering structure of a chemical reaction system is a substructure $\Gamma^* = (\mathcal{M}^*, \mathcal{J}^*)$ of the reaction network $\Gamma = (\mathcal{M}, \mathcal{J})$ that satisfies the following two conditions:

$$\mathcal{J}^* \supseteq \left\{ j \middle| \frac{\partial r_j}{\partial u_m} \neq 0, m \in \mathcal{M}^* \right\}, \tag{9.7a}$$

$$|\mathcal{M}^*| - |\mathcal{J}^*| + N^* = 0, \tag{9.7b}$$

where N^* is the number of independent cycles consisting of only reactions in $\mathcal{J}^*$ (8B Mathematical Method). The condition (9.7) implies that (a) all reactions influenced by chemicals in $\mathcal{M}^*$ are included in $\mathcal{J}^*$ and (b) the number of components in the substructure satisfy the following arithmetic equation: (*number of chemicals*) – (*number of reactions*) + (*number of cycles*) = 0. The buffering structure Γ^* is represented as a square block in matrix $\mathbf{A}$ as follows:

$$\mathbf{A} = \begin{pmatrix} \boxed{\mathbf{A}_{\Gamma^*}} & * \\ \mathbf{0} & \boxed{\mathbf{A}_{\overline{\Gamma^*}}} \end{pmatrix}. \tag{9.8}$$

where $\boxed{\mathbf{A}_{\Gamma^*}}$ is a square block. Let its size be $I \times I$. Then $\boxed{\mathbf{A}_{\overline{\Gamma^*}}}$ is a square block of size $(J - I) \times (J - I)$. Both $\boxed{\mathbf{A}_{\Gamma^*}}$ and $\boxed{\mathbf{A}_{\overline{\Gamma^*}}}$ have at least one nonzero component. Also, $\mathbf{0}$ is a block of size $(J - I) \times I$ such that all components are 0.

In the following, we will explain the essence of the structural bifurcation analysis, by which important aspects of the steady-state bifurcation of a chemical reaction system are determined only from the structural information of the network (Okada et al. 2018, 2021). (1) Factorization: we can study the bifurcation behaviors of a reaction network by decomposing it into smaller subnetworks based on the buffering structures. The bifurcation condition of the entire system is given as the product of the bifurcation conditions of each structure. (2) Inducing parameters: for each substructure, we can identify reactions in the network containing bifurcation parameters that can induce the bifurcation of the substructure. (3) Bifurcating chemicals: for each substructure, we can identify chemicals on the network that exhibit bifurcation behaviors when each substructure satisfies its bifurcation condition.

The properties of equilibrium points in dynamical systems have been discussed using the Jacobian matrix $\boldsymbol{J}$. In particular, it is known that the bifurcation point $\lambda(k^*) = 0$ of the equilibrium point can be identified as the point satisfying the determinant $\det(\boldsymbol{J}(k^*)) = 0$ of the Jacobian matrix $\boldsymbol{J}$. In a reaction system with conserved quantities (see Mathematical Method 8A), there are eigenvalues that are identically equal to 0, but in this case, we are considering the case where $\lambda(k^*) = 0$ depending on the parameter. Okada et al. mathematically proved that the following holds

$$\det\boldsymbol{J} = 0 \Leftrightarrow \det\mathbf{A} = 0. \tag{9.9}$$

That is, the bifurcation of the equilibrium point can be captured by the condition $\det\mathbf{A} = 0$ instead of the condition $\det\boldsymbol{J} = 0$. There are two advantages to using matrix $\mathbf{A}$ to discuss bifurcation of equilibrium points. First, by using matrix $\mathbf{A}$, it is possible to discuss the conditions that the network structure must satisfy in order for bifurcation to occur. This is because each nonzero component of matrix $\mathbf{A}$ corresponds directly to each arrow in the reaction network (Fig. 9.5). Therefore, discussing the bifurcation conditions based on the patten of nonzero components in matrix $\mathbf{A}$ is equivalent to discussing the bifurcation conditions of the network structure. The structure of the Jacobian matrix $\boldsymbol{J}$ may correspond to the dependence between variables, or the structure of the regulatory network, but it has a weak correspondence with the chemical reaction network.

The second advantage of using matrix $\mathbf{A}$ is the significance of the buffering structures in bifurcation analysis. When there is a buffering structure in a system, the bifurcation conditions of the entire system can be factorized based on the buffering structure. As shown in Eq. (9.8), the buffering structure corresponds to a square nonzero block on matrix A. Since $\det\mathbf{A} = \det\mathbf{A}_{\Gamma^*} \times \det\mathbf{A}_{\overline{\Gamma^*}}$ in this case, the bifurcation condition of the entire system, $\det\mathbf{A} = 0$, can be decomposed into two terms: one corresponding to the buffering structure and the other corresponding to the outside of the buffering structure. In other words:

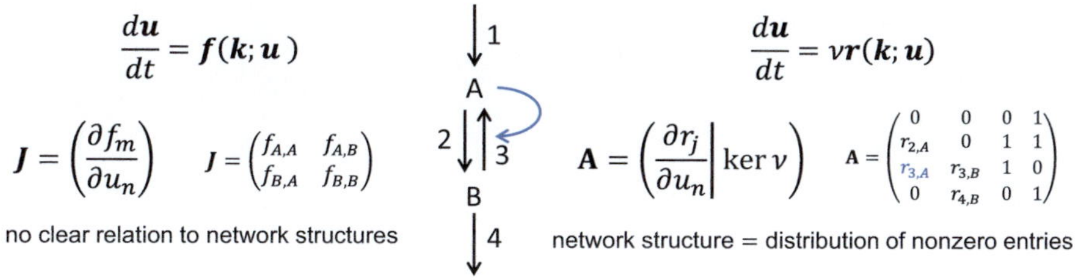

Fig. 9.5 Comparison of Jacobian matrices $\boldsymbol{J}$ and matrix $\mathbf{A}$ for an example chemical reaction network. The black arrows indicate state transitions caused by reactions, and the blue arrows indicate positive regulation from chemical A to reaction 3. In matrix $\mathbf{A}$, there is a one-to-one correspondence between each arrow in the network and each nonzero component of the matrix, except for reaction 1, which is the inflow into the system

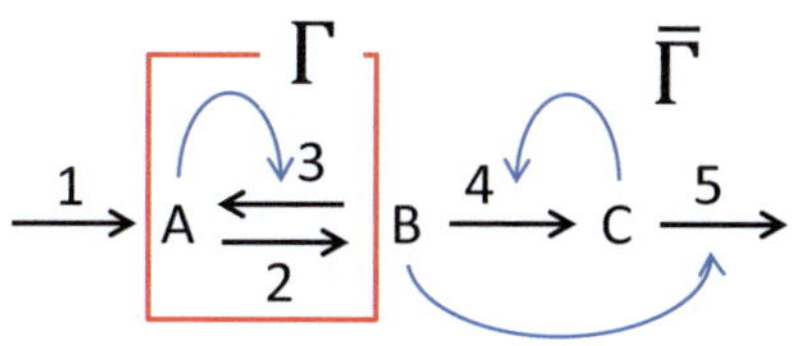

Fig. 9.6 An examples of a chemical reaction network. The black arrows indicate the state transition caused by the reactions, and the blue arrows indicate the positive regulations from the chemicals to the reactions. The red boxes indicate a buffering structure Γ, and $\overline{\Gamma}$ denotes the outside of Γ

$$\begin{gathered}\det\mathbf{A} = \det\mathbf{A}_{\Gamma^*} \times \det\mathbf{A}_{\overline{\Gamma^*}} = 0 \\ \Leftrightarrow \\ \det\mathbf{A}_{\Gamma^*} = 0 \quad \text{or} \quad \det\mathbf{A}_{\overline{\Gamma^*}} = 0\end{gathered} \tag{9.10}$$

Let us take a specific example. Figure 9.6 shows a chemical reaction system containing three chemicals and five reactions. The black arrows indicate reactions, and the blue arrows indicate possible regulation. It is a small system, but possesses one buffering structure ($\Gamma = (\{A\}, \{2, 3\})$, $\chi(\Gamma) = -1 + 2 - 1 + 0 = 0$).

Indeed, the matrix $\mathbf{A}$ of this system, by choosing the order of the rows and columns, becomes

$$\mathbf{A} = \begin{pmatrix} r_{2,A} & 1 & 0 & 0 & 1 \\ r_{3,A} & 1 & r_{3,B} & 0 & 0 \\ 0 & 0 & 0 & 0 & 1 \\ 0 & 0 & r_{4,B} & r_{4,C} & 1 \\ 0 & 0 & r_{5,B} & r_{5,C} & 1 \end{pmatrix} \tag{9.11}$$

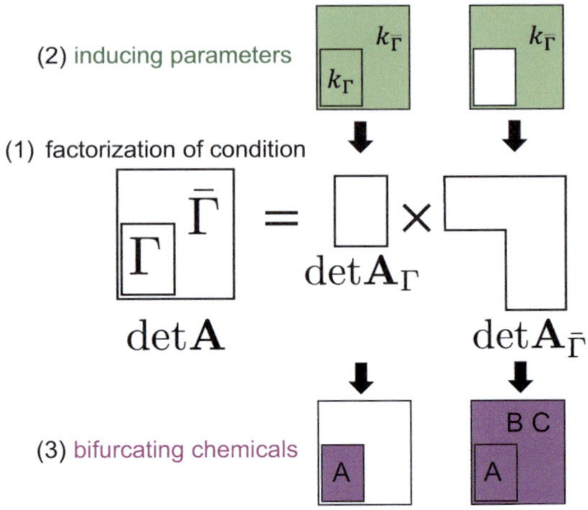

Fig. 9.7 Schematic diagram of structural bifurcation analysis. After the decomposition, (1) bifurcation condition, (2) inducing parameters, and (3) bifurcating chemicals for each substructure can be determined solely from the structure of the chemical reaction network

With a square nonzero block. Here $r_{2,A}$, $r_{3,B}$, $r_{4,B}$, and $r_{5,C}$ correspond to the reactions shown in black in Fig. 9.6, and $r_{3,A}$, $r_{4,C}$, and $r_{5,B}$ correspond to the regulations shown in blue in Fig. 9.6. The steady-state bifurcation conditions for this system are given by

$$\det\mathbf{A} = \det\mathbf{A}_\Gamma \times \det\mathbf{A}_{\bar\Gamma} = (r_{2,A} - r_{3,A})(r_{4,B}r_{5,C} - r_{4,C}r_{5,B}) = 0 \tag{9.12}$$

In the following, structural bifurcation analysis is described in detail from three points (Fig. 9.7).

1. Structural conditions for a steady-state bifurcation

The conditions for this system to produce an equilibrium-point bifurcation are $r_{2,A} - r_{3,A} = 0$ or $r_{4,B}r_{5,C} - r_{4,C}r_{5,B} = 0$. Let's call the former the bifurcation condition for Γ and the latter the bifurcation condition for $\bar\Gamma$. Since all $r_{j,m}$ are positive, in order for the bifurcation condition for Γ to be satisfied, $r_{3,A}$ must not be equal to 0, that is, there must be a regulation from chemical A to reaction 3. Otherwise, the conditional expression will always take a positive value and cannot become 0. Similarly, in order for the bifurcation condition of $\bar\Gamma$ to be satisfied, $r_{4,C}r_{5,B} \neq 0$, meaning that both the regulation from chemical C to reaction 4 and the regulation from chemical B to reaction 5 must be present. In addition, reaction r_1, which does not depend on the concentration of any chemical, does not appear in the components of matrix $\mathbf{A}$, and does not contribute to the bifurcation condition. In this way, for each substructure obtained by network decomposition based on the buffering structure, we can determine the necessary conditions for the network structure to produce the bifurcation and the regulations needed for the necessary conditions.

2. Bifurcation parameters that induce a steady-state bifurcation

Recall that Γ is a buffering structure. It was shown in Chap. 8 that in the steady state of a chemical reaction system, "the effect of a parameter change inside a buffering structure remains inside the buffering structure." This means that the reaction parameters that induce bifurcations for each of Γ and $\overline{\Gamma}$ are also confined to a finite range on the network. In the following discussion, $\widehat{u}_\Gamma$ and $\widehat{u}_{\overline{\Gamma}}$ are the sets of chemical concentrations in Γ and $\overline{\Gamma}$, respectively; $\widehat{r}_\Gamma$ and $\widehat{r}_{\overline{\Gamma}}$ are the sets of reaction rates in Γ and $\overline{\Gamma}$, respectively; k_Γ and $k_{\overline{\Gamma}}$ are the sets of parameters in Γ and $\overline{\Gamma}$, respectively.

Let us focus on the left side of Fig. 9.7, the bifurcation parameter that causes the bifurcation of Γ. k_Γ can be a bifurcation parameter for the bifurcation of Γ, since a change in the parameter k_Γ contained in a reaction within Γ affects the chemical concentration $\widehat{u}_\Gamma$ and the reaction rate $\widehat{r}_\Gamma$ in Γ. In addition, a change in parameter $k_{\overline{\Gamma}} = \{k_4, k_5\}$ contained in a reaction within $\overline{\Gamma}$ (ouside Γ) also affects the chemical concentration $\widehat{u}_\Gamma \cup \widehat{u}_{\overline{\Gamma}} = \{\widehat{u}_A, \widehat{u}_B, \widehat{u}_C\}$ and the reaction rate $\widehat{r}_\Gamma \cup \widehat{r}_{\overline{\Gamma}} = \{\widehat{r}_1, \widehat{r}_2, \widehat{r}_3, \widehat{r}_4, \widehat{r}_5\}$ of both Γ and $\overline{\Gamma}$. The bifurcation condition of Γ includes only the reaction rates $\widehat{r}_\Gamma = \{\widehat{r}_2, \widehat{r}_3\}$ in Γ, which depend only on the chemical concentration $\widehat{u}_\Gamma$ in Γ, but since $\widehat{u}_\Gamma$ depends on the reaction parameter $k_{\overline{\Gamma}}$ within $\overline{\Gamma}$, $k_{\overline{\Gamma}}$ can be a bifurcation parameter for a bifurcation of Γ. In summary, it is $k_\Gamma \cup k_{\overline{\Gamma}}$, i.e., the reaction parameters contained in r_1, r_2, r_3, r_4, and r_5, that can be the bifurcation parameters for the bifurcations of Γ. (Where r_1 does not appear in the bifurcation condition as described above, so the parameters in r_1 have no effect on bifurcation.)

Next, let us consider the right side of Fig. 9.7, the bifurcation parameter that induces a bifurcation of $\overline{\Gamma}$. $k_{\overline{\Gamma}}$ can be a bifurcation parameter for a bifurcation in $\overline{\Gamma}$, since a change in the parameter $k_{\overline{\Gamma}}$ contained in the reactions in $\overline{\Gamma}$ affects the chemical concentration $\widehat{u}_{\overline{\Gamma}}$ and the reaction rate $\widehat{r}_{\overline{\Gamma}}$. On the other hand, a change in the parameter $k_\Gamma = \{k_2, k_3\}$ contained in the reactions in Γ does not affect the chemical concentration $\widehat{u}_{\overline{\Gamma}} = \{\widehat{u}_B, \widehat{u}_C\}$ and the reaction rate $\widehat{r}_{\overline{\Gamma}} = \{\widehat{r}_1, \widehat{r}_4, \widehat{r}_5\}$ in $\overline{\Gamma}$. Since the bifurcation condition of $\overline{\Gamma}$ includes only the reaction rates $\widehat{r}_{\overline{\Gamma}}$ in $\overline{\Gamma}$, and they depend only on the chemical concentration $\widehat{u}_{\overline{\Gamma}}$ in $\overline{\Gamma}$, the reaction parameters k_Γ in Γ cannot be the bifurcation parameters for the bifurcation of k_Γ. Only $k_{\overline{\Gamma}}$, the reaction parameters in r_1, r_4, and r_5 can be bifurcation parameters that induce bifurcation of $\overline{\Gamma}$. (The parameter in r_1 has no effect on bifurcation.)

3. Bifurcating Chemicals

When the bifurcation condition for each substructure discussed in (1) is satisfied, the chemicals that exhibit bifurcation behavior can be specified on the network. This is determined by the nonzero component (called support) of the eigenvector corresponding to eigenvalue 0 of matrix $\mathbf{A}$. When $\det\mathbf{A}_\Gamma = 0$, that is, the bifurcation condition for Γ is satisfied, the eigenvector corresponding to the zero eigenvalue has nonzero components (support) only on the chemicals $\widehat{u}_\Gamma$ in Γ. On the other hand, when $\det\mathbf{A}_{\overline{\Gamma}} = 0$, i.e., the bifurcation condition for $\overline{\Gamma}$ is satisfied, the zero vector has the support on the chemicals $\widehat{u}_\Gamma \cup \widehat{u}_{\overline{\Gamma}}$ of both Γ and $\overline{\Gamma}$. The details of the mathematics and proofs are given in the references (Okada et al. 2018, 2021). In other words, when the bifurcation condition for Γ is satisfied, the only chemical that exhibits bifurcation behavior is A. When the bifurcation condition for $\overline{\Gamma}$ is satisfied, the chemicals that exhibit bifurcation behavior are all A, B, and C.

9.4 Numerical Examples to Verify Structural Bifurcation Analysis

Up to this point, the discussion has proceeded based only on the network structure, without giving specific forms of the reaction rates r_j or its parameter values. We analyzed numerically whether these behaviors predicted based on the network structure are actually observed by giving specific reaction rate functions r_j to the system (Fig. 9.8). We easily found parameter conditions satisfying $\det\mathbf{A}_\Gamma = 0$ and $\det\mathbf{A}_{\overline{\Gamma}} = 0$. These should correspond to the bifurcation points of Γ and $\overline{\Gamma}$, respectively. Especially for the latter, changing the parameters in r_2 and r_3 should not affect the bifurcation behaviors. In addition, the chemicals that exhibit bifurcation behavior should be only molecule A at the bifurcation of Γ, and all of chemicals A, B, and C at the bifurcation of $\overline{\Gamma}$. As Fig. 9.8 shows, all of these predicted behaviors were actually observed in the numerical analysis. Note that in the bifurcation of Γ (Fig. 9.8, left), the upstream chemical A changes its value drastically, while the downstream chemicals B and C do not change their values at all. It is also important to note that these are the only two possible patterns of bifurcation that the system can exhibit. In other words, it is generally impossible for only molecule B, only molecule C, or molecules A and B to exhibit bifurcation behaviors.

Structural bifurcation analysis is even more effective when the reaction system has multiple buffering structures (Fig. 9.9). The bifurcation condition of the entire system can be factorized into many terms corresponding to buffering structures. For each term, a range of bifurcation parameters can be specified on the network, and chemicals exhibiting bifurcation behavior can be specified on the network.

Finally, an example of this analysis applied to the reaction network of the central metabolic system of *E. coli* is shown in Fig. 9.10. This system contains 28 chemicals and 46 reactions and is of a size and complexity that would make bifurcation analysis impossible with the conventional understanding in mathematical science. The reaction network is the same as that analyzed in Chap. 8. Since there are 17 buffering structures, the bifurcation condition is also described as the product of 17 terms. Examination of each of the 17 conditional equations shows that 16 of them are always positive or negative and cannot take the value of 0. In other words, in this network, bifurcation occurs only when a single subsructure satisfies its bifurcation condition $\det\mathbf{A}_{\mathrm{red}} = 0$. The substructure is shown in red in the figure. In addition, the bifurcation parameter for the bifurcation of the red substructure is the parameters included in the reactions indicated by red and green only. Furthermore, when the red

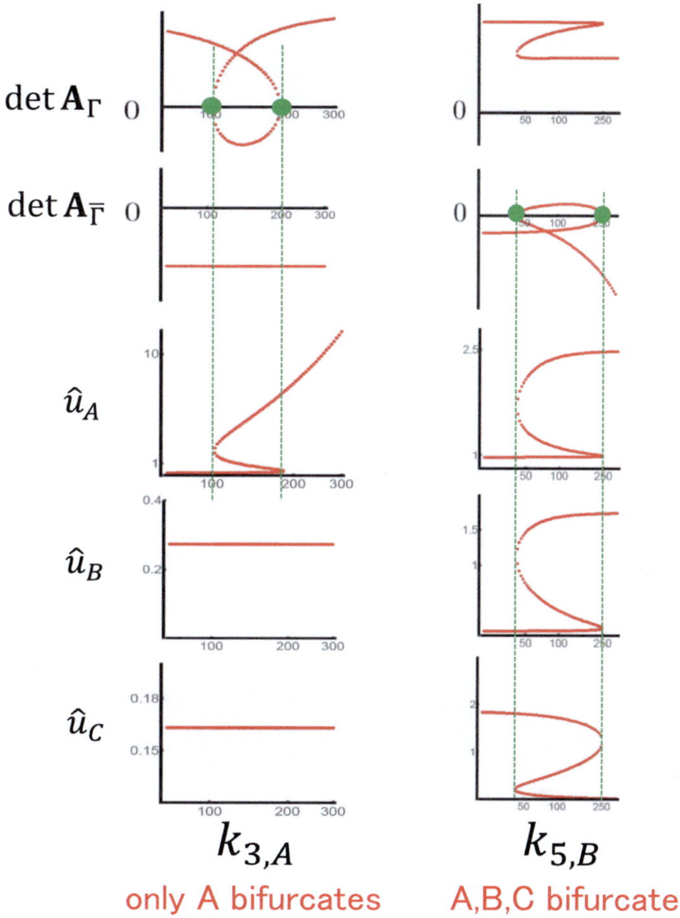

Fig. 9.8 Verification of the structural bifurcation analysis by numerical analysis. det$\mathbf{A}_\Gamma$, det$\mathbf{A}_{\bar\Gamma}$, and steady state $\hat{\boldsymbol{u}}$ were computed numerically for the network in Fig. 9.6, given specific reaction rate functions. Left: Bifurcation for a change in the constant $k_{3,\,A}$ in r_3; Right: Bifurcation for a change in the constant $k_{5,\,B}$ in r_5

substructure satisfies the bifurcation condition det$\mathbf{A}_{\text{red}} = 0$, the only chemicals that exhibit bifurcation behavior are the chemicals indicated in red and blue.

These theoretical predictions, determined by the network structure, were confirmed by numerical analysis. As shown in the lower part of Fig. 9.10, bifurcation behavior was indeed observed by changing the parameters indicated in red and green. In this case, only the chemicals marked in red and blue exhibited bifurcation behavior. Note that it is the only possible bifurcation pattern of this system. This fact is determined by structural bifurcation analysis at the point when the network structure is given, even before the reaction rate function is given.

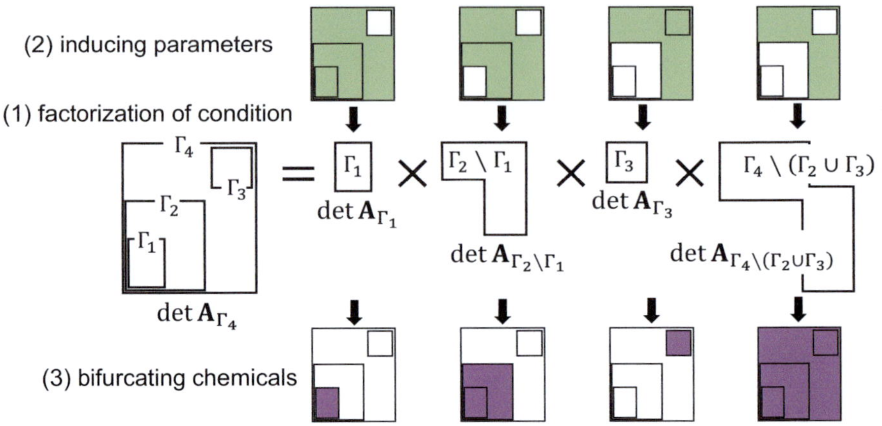

Fig. 9.9 Schematic of structural bifurcation analysis when the reaction system has multiple buffering structures

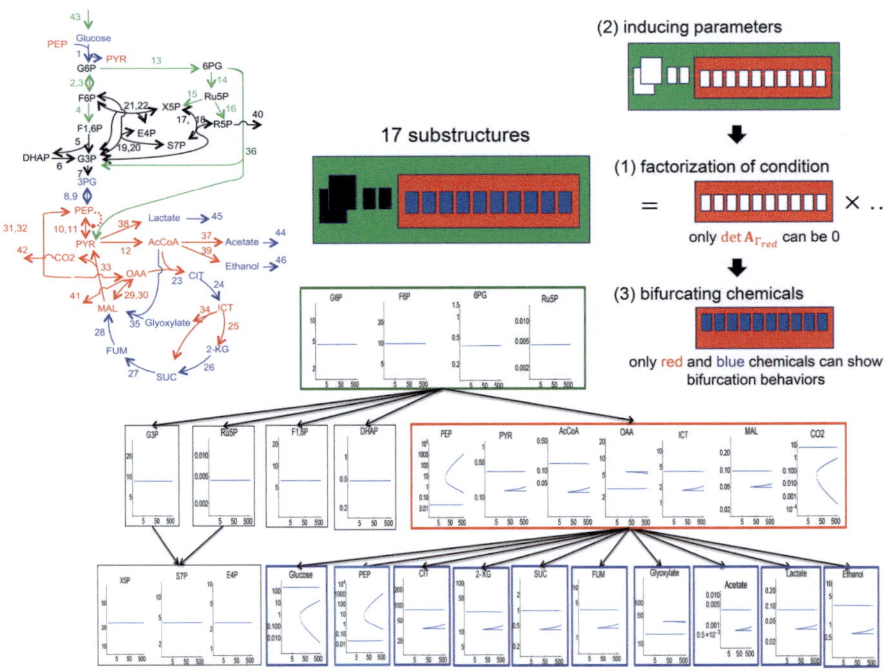

Fig. 9.10 Application of structural bifurcation analysis to the *E. coli* central metabolic system. (1) Bifurcation condition: The bifurcation condition of the entire system is decomposed into 17 terms (corresponding to 17 substructures), of which only one (the red part in the figure) can cause bifurcation. (2) Bifurcation parameter: The bifurcation of this (red) substructure is induced only by the reaction parameters contained in the substructure (red) and the glycolysis system (green). (3) Bifurcation behavior molecules: When the bifurcation condition of red substructure is satisfied, the bifurcation behavior is observed only in the concentration of chemicals in the substructure (red) and the TCA circuit (blue)

9.5 Summary

In this chapter, examples of bifurcation phenomena observed in living systems and the fundamentals of bifurcation analysis were explained. I also introduce the latest theory based solely on network structure, structural bifurcation analysis. In the mathematical sciences, the importance of bifurcation has been well understood and has attracted the attention of mathematical scientists. Although bifurcation analysis is not yet common in the life sciences, its importance is expected to be increasingly understood in the future. The structural bifurcation analysis presented here has the great potential to promote the situation. This method enables qualitative analysis of bifurcation in chemical reaction systems based solely on the network structure. It is expected to make a significant contribution to the understanding of the behavior of complex biological systems.

References

Borisuk MT, Tyson JJ (1998) Bifurcation analysis of a model of mitotic control in frog eggs. J Theor Biol 195:69–85

Mochizuki A, Fiedler B (2015) Sensitivity of chemical reaction networks: a structural approach. 1. Examples and the carbon metabolic network. J Theor Biol 367:189–202

Novak B, Tyson JJ (1993) Numerical analysis of a comprehensive model of M-phase control in Xenopus oocyte extracts and intact embryos. J Cell Sci 106:1153–1168

Okada T, Mochizuki A (2016) Law of localization in chemical reaction networks. Phys Rev Lett 117:048101

Okada T, Mochizuki A (2017) Sensitivity and network topology in chemical reaction systems. Phys Rev E 96:022322

Okada T, Tsai J-C, Mochizuki A (2018) Structural bifurcation analysis in chemical reaction networks. Phys Rev E 98:012417

Okada T, Mochizuki A, Furuta M, Tsai J-C (2021) Flux-augmented bifurcation analysis in chemical reaction network systems. Phys Rev E 103:062212

Ozbudak EM, Thattai M, Lim HN, Shraiman BI, van Oudenaarden A (2004) Multistability in the lactose utilization network of Escherichia coli. Nature 427:737–740

MIX
Papier aus verantwortungsvollen Quellen
Paper from responsible sources
FSC® C105338

If you have any concerns about our products,
you can contact us on
ProductSafety@springernature.com

In case Publisher is established outside the EU,
the EU authorized representative is:
Springer Nature Customer Service Center GmbH
Europaplatz 3, 69115 Heidelberg, Germany

Printed by Libri Plureos GmbH
in Hamburg, Germany